PROCESS SAFETY MANAGEMENT

Ian S. Sutton

First Edition

Southwestern Books, Houston, Texas, United States of America

PROCESS SAFETY MANAGEMENT

Copyright © 1997 by Southwestern Books

ISBN 1-57502-528-0
Library of Congress Cataloging-in-Publication Data
97-92248

Sutton, Ian S.
    Process Safety Management/
    Ian S. Sutton

Published and distributed by Southwestern Books, Houston, Texas

Southwestern Books, 2437 Bay Area Blvd., Ste. 195, Houston,
Texas, United States of America, 77058

phone:	281-488-7767
fax:	281-488-2259
e-mail:	swbooks@iwl.net
web:	www.southwestern-books.com

Printed in the USA by

MORRIS PUBLISHING
3212 East Highway 30 • Kearney, NE 68847 • 1-800-650-7888

WARNING — DISCLAIMER

ADDITIONAL MATERIALS

Additional titles available from Southwestern Books include:

Writing Operating Procedures For Process Plants
Audit Protocols For Process Safety (1998)
A Practical Guide to Statistical Quality Improvement
How To Write And Publish A Professional Book (1998)

Training Courses on the materials in these books are available. For additional information, please contact Southwestern Books,

ABOUT THE AUTHOR

Ian S. Sutton is a chemical engineer with over 25 years of diversified experience in the process industries. He is a principal engineer with Fluor Daniel Inc. in Houston, Texas, where he specializes in Process Safety Management. He is a graduate of the University of Nottingham, U.K. and is a professional engineer registered in the State of Texas.

CONTENTS

Chapter 3 — Employee Participation

Chapter 4 — Process Safety Information

Chapter 8 — Contractors

Chapter 9 — Prestartup Safety Review

Chapter 10 — Mechanical Integrity

Chapter 11 — Hot Work

Chapter 12 — Management Of Change

Chapter 13 — Incident Investigation

Chapter 16 — Quantification Of Risk

Figures

PREFACE

I have been a process safety management consultant for the last ten years or so. During that time, I have been privileged to have had the opportunity to work with a wide variety of clients in many industries and areas of technology. During the last five years, most of my effort has been directed toward helping clients meet the requirements of the OSHA Process Safety Management regulation, 29 CFR 1910.119 and other regulatory standards. In order to capture some of the knowledge that we as an industry and that I as an individual have gained, I decided to write a series of books on the topic of process safety management. As Trevor Kletz said in his book <u>Why Accidents Happen</u>, organizations do not have memories, only people do. Therefore, those of us who have accumulated experience in any field can do a service to those just entering the business by writing down what we know and have learned.

The first book in this series, which was published in late 1995, was <u>Writing Operating Procedures For Process Plants</u>. The feedback that I received on it was rewarding and gratifying. Although writing a book is very hard work, I felt that it was worth the effort to know that I was able to help others with their process safety projects. The second book in the series — this one — covers all aspects of process safety, and so is called <u>Process Safety Management</u>. Writing this one has been even harder work than was <u>Writing Operating Procedures</u>, probably because the topic is so broad in scope. Anyway, I hope that you the reader find this book as useful as its predecessor appears to have been. Since I seem to be a glutton for punishment, I am already well into the third book in the series, which is called <u>Audit Protocols For Process Safety</u>. It will contain a series of audit questions that people can use when they are reviewing process safety programs. The book will also contain checklist-style questions which can be used to create customized audit protocols.

I am also working on another book which I hope will help other professionals organize and write down their knowledge and experience. Its title is <u>Writing And Publishing A Technical Book</u>. In it, I describe the business of writing and publishing a professional book, and what it takes to self-publish your own work.

The manuscript of <u>Process Safety Management</u> was reviewed not only by my colleagues at Fluor Daniel, but also by engineers and safety specialists working for a wide variety of companies in many areas of technology. All comments and advice have been truly invaluable, and are deeply appreciated. Naturally, the responsibility for errors or oversights is mine. I would particularly like to express my thanks and appreciation to Raymond E. Brandes, who provided much of the understanding that I gained when I was a process safety neophyte. Many of the ideas and insights that are presented here came from him. Indeed, I suspect that some of the ideas that I now firmly believe to be mine really originated with Ray.

If you would like to be a reviewer for future books (such as <u>Audit Protocols For Process Safety</u>) please let us know. All that we can offer in return is an opportunity to read the book some months before it is published and a complimentary copy of the final product. But many of the reviewers of this book appear to have enjoyed helping out.

I have tried to make this book as practical as possible, and to base my writing on actual work that I and others have carried out. Nevertheless, all the examples used in this book are fictional; they are based on real-life situations, but the details have been altered so as to ensure that no company, process or individual can be identified.

I have tried to use inclusive language at all times; for example "he or she" instead of just "he." Unfortunately, the English language does not have a neutral word for a single person, and the "he or she" construct is rather awkward, and I don't like using it. I considered following Trevor Kletz' example by just using the masculine tense, but decided at least to make an attempt to be inclusive unless it made the text just too clumsy or ugly.

Should you wish to get in contact with us, we can be reached at swbooks@iwl.net or www.southwestern-books.com. Our phone number is 281-488-7767, and our fax is 281-488-2259.

As always, my appreciation to Val, Ann and Peter for their patience and understanding. And now, using the words of Edmund Spenser once more: "Goe little book: thy selfe present."

Chapter 1

PRINCIPLES OF PROCESS SAFETY MANAGEMENT

INTRODUCTION

 No process plant is designed or operated to have accidents, yet accidents occur all too frequently. The root cause in all cases is that operating conditions deviated outside the permissible range, resulting in an uncontrolled release of chemicals and/or energy. Process Safety Management (PSM) is a management system that helps everyone working on the plant — managers, engineers, operators, maintenance personnel and contract workers — to achieve a safe operation by defining safe operating limits, and then making sure that the process stays within those limits. Hence, PSM is concerned with three fundamental issues.

1. *Defining Control*. A person running a process must know what the safe control limits are, and whether the plant is operating within them. For example, the safe temperature range for a certain reaction may be 125-145°C. If the actual temperature deviates outside of that range, then that reaction is — by definition — out of control and potentially unsafe, and action must be taken to bring the temperature back into the correct range.

   When a plant is new, the control points are defined by its designers. As operating experience is accumulated, new control points will be implemented, using the Management of Change process.

   This concept of quantified control limits applies to more than just control system information such as reaction temperatures. It covers all aspects of a plant's performance, including operations, engineering, maintenance and administration. For example, it may be required that the wall thickness of a certain section of pipe be greater than 10 mm. If the thickness falls below this value say due to corrosion, then the system is out of control because the piping may fail and cause a leak of hazardous chemicals.

2. *Attaining Control.* Once a control range has been defined, management must determine how to operate their plant so that it stays within that range. In the case of the reaction temperature example, instruments must be adjusted and operators trained so as to achieve the 125-145°C range. For the pipe wall thickness example, process conditions must be such as to ensure that corrosive materials cannot enter the pipe.

3. *Response To Loss Of Control.* Once the defined control range has been determined and achieved, all the people involved in running or maintaining the unit must know how to identify an out-of-control situation, what its consequences might be, and how they should respond to it.

Having provided this philosophical background to PSM, it is useful to analyze the phrase *Process Safety Management* by examining its component words.

The first word in the phrase is *Process.* PSM is concerned with process issues (such as reactor temperatures or pipe thickness) as distinct from *occupational* safety issues, such as trips and falls.

The second word is *Safety.* The driving force for most PSM programs is the need to meet a safety regulation. However, the benefits of process safety go beyond this. If management is in full control of a facility, all aspects of the operations will improve. Not only is a properly controlled plant safer, it is also more profitable and it is environmentally cleaner. Hence, the justification for process safety extends beyond simply complying with a regulation, or even improving safety; process safety makes sense on its own terms.

The third and final word in the phrase PSM is *Management.* (In this context, a manager is taken to be anyone who has some degree of control and independence — this would include most operators, engineers and maintenance workers.) Control can only be achieved through good management practices. This is why topics such as Root

Cause Analysis are so important. It is vital that problems be controlled by improving management systems.

Process safety management is non-prescriptive. This means that the regulations and other standards in this field generally provide very little detail. Basically, they say "do whatever it takes *on your facility* not to have accidents." It is up to the managers and operators to determine what the optimum operation should be. What is appropriate in one location may or may not be appropriate in another. The PSM standards simply require that programs be in place, and that they be adhered to. (In this regard, PSM is similar to ISO 9000 and other quality standards, which also require that companies set their own standards, and then adhere to them.) In practice, although there are no universally "correct answers" as to what needs to be done to achieve a safe operation, it is important not to exaggerate the differences between different facilities and companies. Many operations, such as starting a pump or training a contract worker, are quite similar from site to site and from company to company. Therefore, it is possible to develop high quality, generic systems that can be used in a wide variety of situations. This saves time and money, and improves quality.

In addition to being non-prescriptive, PSM is performance-based. This means that the only true measure of success is not to have accidents. One consequence of this insight is that it not possible, at least from a theoretical point of view, ever to achieve "compliance." The only truly acceptable level of safety is zero accidents. Yet, no matter how well run a facility may be, this is a theoretically unattainable goal. In spite of the fact that many companies set a target goal of "zero accidents", risk can never be zero and accidents can always happen. Indeed, if a unit operates for long enough, it is *certain* —statistically speaking— that there will be an accident. Yet, given that real-world targets are needed, someone in authority has to set a level for "acceptable safety." This they are very reluctant to do — in particular regulatory agencies will never place a numerical value on human life and suffering; any number that they develop would inevitably generate a lot of controversy. Yet somehow, working targets have to be provided, otherwise the facility personnel do not know what they are shooting for.

One way of resolving this dilemma is to view PSM as being a *process* that can be divided into *programs*. Being a process, PSM can never be complete — there are always ways of improving safety. However, the programs within PSM can be finite in size, and can have clear and acceptable definitions for the word "completion."

For example, a company may choose to divide its initial process safety program into three phases. During Phase One, the company will conduct a baseline audit and determine what its goals should be to meet both regulatory and non-regulatory targets. Following this, Phase Two will focus on the company's process safety information base, which provides the essential foundation for the other elements of process safety. Once this information base is in place, Phase Three will be concerned with conducting hazards analyses on all units, and on making sure that all operating procedures are complete, accurate and up-to-date. Phases Four and onward follow on. By using a program such as this, management will be able to measure progress with regards to process safety, but no one will ever be under the impression that the program is completely finished.

A term that is used in this context is "good engineering practice." This is a very important concept for a non-prescriptive regulation, such as OSHA's PSM, and it can be defined in one of two ways. First, guidance can be obtained from authoritative sources, such as books and journals published by professional societies (the books published by the Center For Chemical Process Safety are a good example.) Second, a company may choose to develop a standard internally and then to declare it "good engineering practice", in its judgment. Should they do this, they would have to be able to document how they arrived at their opinion, and to defend it if challenged.

It is likely that a clear understanding of the term "good engineering practice" will become increasingly important as process safety programs mature. If there is an accident on which process safety work (such as a process hazard analysis or a rewriting of the operating procedures) has been performed, then the resulting investigation will have to determine if the work was performed to a high enough standard. In one regard, it could be argued *ipso facto* that the process safety work was not of sufficiently high standard because the accident took place. Yet, safety can never be perfect and risk can never be zero, hence there is always a chance of an accident no matter how well the process safety work was done. Consequently, it is likely that outside experts will be called in to determine if the work meets the standard of "good engineering practice."

ORGANIZATION OF THIS BOOK

The purpose of this book is to assist companies with the development and implementation of their process safety programs. The book is structured around the United States OSHA process safety management standard because so many companies have to follow it. Many international companies have chosen to meet the requirements of the OSHA standard world-wide as an expression of their commitment as global citizens. Even if a company does not have to follow the OSHA standard, it still makes sense to structure a process safety program around it because it is organized in a manner very similar to that of other industry guidelines and protocols in this area.

This book is divided in three principal sections. The first section — Chapters 1 and 2 — provides an overview of process safety and of the pertinent regulations in the United States. The second section — Chapters 3 through 15 — describes each of the elements of PSM (as defined by OSHA), and provides guidance as to how they can be implemented. Each of the chapters in the second section has the same general structure, as shown in Table 1-1. (The quotations from OSHA are direct; neither their spelling nor grammar has been modified.)

Table 1-1

Chapter Contents

Introduction The OSHA Regulation OSHA Guidance Analysis Of The Regulation Details . . .

The third and final section of the book — Chapters 16 and 17 — discusses the use of quantitative risk management techniques and the practical implementation of a process safety program.

DEFINITION OF PROCESS SAFETY MANAGEMENT

Based on the above discussions, the following definition is offered for the term Process Safety Management.

Process Safety Management is a process, involving all managers, employees and contract workers, that aims to minimize uncontrolled change from design and/or operating intent.

An alternative definition of Process Safety Management provided by the Center for Chemical Process Safety (CCPS) is:

The application of management systems to the identification, understanding, and control of process hazards to prevent process-related injuries and incidents.

For most companies in the United States, the driving force for process safety during the 1980's and 90's has been regulatory. In particular, the OSHA Process Safety Management Standard (29 CFR 1910.119), which was promulgated in 1992. This standard required almost all companies in the United States that process chemicals to implement a PSM program. The OSHA standard also requires that PSM programs be complete by mid-1997.

Although OSHA does not define the term Process Safety Management, it does offer the following guidance with respect to the 29 CFR 1910.119 regulation.

The major objective of process safety management of highly hazardous chemicals is to prevent unwanted releases of hazardous chemicals especially into locations which could expose employees and others to serious hazards. An effective process safety management program requires a systematic approach to evaluating the whole process. Using this approach the process design, process technology, operational and maintenance activities and procedures, non routine activities and procedures, emergency preparedness plans and procedures, training programs, and other elements which impact the process are all considered in the evaluation. The various lines of defense that have been incorporated into the design and operation of the process to prevent or mitigate the release of hazardous chemicals need to be evaluated and strengthened to assure their effectiveness at each level. Process safety management is the proactive identification, evaluation and mitigation or prevention of chemical releases that could occur as a result of failures in process, procedures or equipment.

The process safety management standard targets highly hazardous chemicals that have the potential to cause a catastrophic incident. This standard as a whole is to aid employers in their efforts to prevent or mitigate episodic chemical releases that could lead to a catastrophe in the workplace and possibly to the surrounding community. To control these types of hazards, employers need to develop the necessary expertise, experiences, judgment and proactive initiative within their work force to properly implement and maintain an effective process safety management program as envisioned in the OSHA standard.

PROCESS SAFETY REGULATIONS IN THE UNITED STATES

In the United States, there are two federal regulations that cover process safety: OSHA's 29 CFR 1910.119 (already referred to) and EPA's 40 CFR 68. OSHA's mandate is to protect the safety and health of workers, whereas the EPA is more concerned with protection of the public and the environment. There is a good deal of overlap between the two because an accident that hurts a worker could also affect the general public and *vice versa*. Because of these similarities both agencies involved have worked to minimize duplication between their respective standards. The remaining differences tend to reflect the different goals. For example, EPA is more concerned about the consequences of catastrophic releases of toxic materials, whereas OSHA focuses more on activities "inside the fence." Hence EPA emphasizes the need for vapor cloud modeling; whereas OSHA is much less concerned with it.

In addition to these federal regulations, there are some state and local regulations covering process safety. There are also industry standards, which may not have the force of law, but which nevertheless provide important guidance, and can possess considerable authority.

Table 1-2 lists major state and federal process safety regulations in the United States in the chronological order in which they were promulgated.

Table 1-2

Process Safety Regulations (United States)

Region	Year	Title
California	1988	Risk Management Prevention Program (RMPP). Assembly Bill 3777 Article 2, Section 25531, et seq. of Chapter 6.95 of the Health and Safety Code.
New Jersey	1989	Toxic Catastrophe Prevention Act (TCPA) N.J.S.A. 13:1K-19 et seq.
Delaware	1989	Extremely Hazardous Substances Risk Management. Delaware Code, Title 7, Chapter 77
Nevada	1991	Nevada Senate Bill No. 641
United States and Territories	1992	Process Safety Management Of Highly Hazardous Chemicals. Occupational Safety & Health Administration (OSHA), 29 CFR 1910.119.
United States and Territories	1996	Risk Management Program (RMP). Environmental Protection Agency (EPA), 40 CFR 68.

The first regulation to be applied nationally was the OSHA PSM standard in 1992. Consequently it is this standard that has drawn most attention, and that has been the focus of most of the work since then. Its development started in the late 1980's. At that time, there had been a number of serious process plant accidents in the United States. Industry executives from various chemical companies recognized that the introduction of a process safety regulation in the United States was inevitable. Therefore, senior managers from these companies, working with the Organization of Resource Counselors (ORC), set about developing standards that would serve both society's and industry's needs. One of the most important consequences of having a standard developed by industry was that the resulting regulations were non-prescriptive and performance-based. The executives who drafted the standard were trying to avoid the problem of having a large number of lengthy, highly prescriptive, detailed regulations such as are to be found in the environmental and nuclear power businesses. The development of such regulations for the process industries would have been very demanding due to the wide variety of processes and technologies.

The document that this industry group published became the basis of the following regulations and standards:

- OSHA's PSM regulation, 29 CFR 1910.119
- EPA's Risk Management Program, 40 CFR 68
- The American Petroleum Institute's (API) Recommended Practice 750

Although there are differences between these programs, the technical requirements are generally similar. This means that if a company develops a PSM program to meet one standard, it is likely that it has gone most of the way toward meeting the requirements of the others. (Non-technical areas, such as community communication requirements, reporting procedures and the lists of covered chemicals can differ considerably from standard to standard.)

Although process safety management regulations are new, process safety management principles are not. All that the regulations have done is to formalize a process that all companies have always followed to some degree. For example, employee training is an important part of all PSM standards. Yet all companies have always provided some training to their workers. All that has changed is that formal, documented training that consistently covers all employees and that leads to an acceptable level of safety, be implemented. Indeed, if a company claims that starting a process safety program was expensive, both in terms of money and in the time of their key people, a reasonable response is that process safety is not new, and what they are doing now is playing catch up on all the items that have been neglected. This is also fair to those companies that have always had a commitment to process safety. They naturally feel resentment toward competitors who can cut corners and so gain a (short-term) market advantage.

PROCESS SAFETY MANAGEMENT PRINCIPLES

Based on the discussions up to this point, it is clear that some of the most important principles behind a PSM program are as follows.

- *Participation.* PSM is not a management program that is handed down by management to their employees and contract workers; it is a program that involves everyone. The key word is *participation*, which is much more than just *communication*. *All* managers, employees and contract workers are responsible for the successful implementation of PSM. Management must organize and lead the initial effort, but the employees must be fully involved in its implementation and improvement because they are the people who know the most about how a process really operates, and they are the ones who have to implement recommendations and changes. Specialist groups, such as staff organizations and consultants can provide help in specific areas, but PSM is fundamentally a line responsibility.

- *Performance Based.* The measure of success of a PSM program is that there are no incidents. The regulations do not specify how to do this. They merely say that companies should determine what needs to be done in each of their facilities, and then do it. This is why the OSHA standard is so short; the technical section of the regulation occupies only about ten pages.

- *Quantification.* An effective process safety program requires that progress be measured quantitatively. There must be objective goals and tools must be provided for determining how much progress has been made toward achieving those goals. Not only should control variables be quantified (as in the reactor temperature example provided earlier), the PSM program itself should be measured quantitatively with respect to its own goals and objectives (although some elements of PSM, such as Employee Participation, can be difficult to quantify and measure.)

- *Auditing.* An essential part of any management program is the review or audit step. If the program is to work, it is necessary that there be continual feedback as to how it is progressing. This feedback can range all the way from informal reviews to formal audits.

- *Thoroughness.* PSM regulations require thoroughness. For example, a company may have a good training program, but one person may have missed part of it because he or she was off work sick. In order to comply with the PSM requirements, the company will have to make sure that this person is trained, and this his or her personnel files are updated appropriately.

- *On-Going.* PSM is an on-going activity that never ends; it is a process, not a project. Because risk can never be zero there must always be ways of improving safety and operability. Also, regardless of the risk goal, there is still a need for continuous improvement just to retain control.

- *Documentation.* Regulations and standards require that good written records be maintained so that outside auditors can evaluate the status of the system. Keeping good records is not just "bureaucracy." Records provide the foundation for sound decision-making in all areas of plant management.

TYPES OF SAFETY

Earlier in this chapter, a distinction was made between *Process* safety and *Occupational* safety. Process safety, which is the topic of this book, is concerned with process-oriented issues such as runaway chemical reactions, corrosion and the inadvertent mixing of hazardous chemicals. Occupational safety, sometimes referred to as "hard-hat" safety, has to do with normal day-to-day hazards. It covers topics such as vessel entry, vehicle movement and tripping hazards. The relationship between these types of safety is shown in Figure 1-1, which shows that both occupational and process safety are parts of overall *System* safety.

<u>Figure 1-1</u>

<u>Types Of Safety</u>

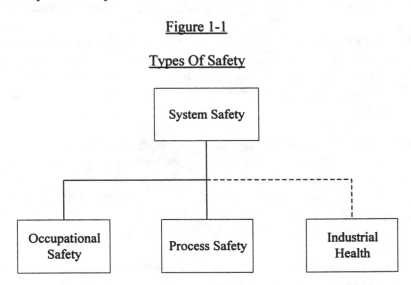

In practice, there is usually considerable overlap between process safety and occupational safety. Indeed, the OSHA PSM regulations themselves incorporate some aspects of occupational safety into process safety. For example, the topic of Human Factors would normally be considered to be part of occupational safety, yet it is referred to in the Process Hazards Analysis section of 29 CFR 1910.119. In spite of this type of overlap, it is useful to keep the two types of safety separate because, if the process safety program starts to incorporate many occupational safety issues, its costs will rise and its schedules will slip.

Figure 1-1 also shows the topic of Industrial Health. Health and safety have a good deal in common and are often coordinated by the same people. Safety is primarily concerned with sudden, catastrophic incidents that could result in serious injury or death, whereas industrial health is more concerned with long-term, chronic problems. Health information can be used when there is a lack of safety data. For example, typically not much is known about the impact of high concentrations of toxic chemicals on the human body over a short period of time. In these situations, safety professionals can extrapolate industrial health data, in which the impact of the same chemicals in low concentrations over longer periods of time is understood. Such an extrapolation may be very approximate, but it does provide some guidance for safety work.

Process safety management does not address malicious acts such as war or terrorism. Nor is it within the scope of PSM to consider how to protect the plant against a knowledgeable employee who decides to sabotage a unit with which he or she is very familiar. Such situations should be left to trained investigators.

HOLISTIC NATURE OF PROCESS SAFETY MANAGEMENT

PSM programs typically include about a dozen major elements. The OSHA standard, the one that most companies in the United States follow, contains fourteen, as listed in Table 1-3.

Table 1-3

Elements Of Process Safety Management

1. Employee Participation
2. Process Safety Information
3. Process Hazards Analysis
4. Operating Procedures
5. Training

6. Contractors
7. Prestartup Safety Review
8. Mechanical Integrity
9. Hot Work
10. Management of Change
11. Incident Investigation
12. Emergency Planning And Response
13. Compliance Audits
14. Trade Secrets

In order to illustrate the problems that can occur when these elements are not properly controlled, a list of some of the major process accidents that have occurred since 1974 is provided in Table 1-4. An approximate figure for the number of fatalities and injuries associated with each is also provided. In each case, one or more of the PSM elements that were a root cause of the accident are identified.

Table 1-4

Some Major Process Accidents

Year	Location	Brief Description
1974	Flixborough, England	Rupture of a temporary pipe led to large release of cyclohexane gas, followed by a massive explosion. 25 deaths. *Management Of Change*
1976	Seveso, Italy	Release of highly potent toxin, TCDD. Approximately 250 injuries. *Process Safety Information*

Year	Location	Brief Description
1984	Bhopal, India	Addition of water to a tank containing a hazardous chemical led to a release of isocyanate vapors. Approximately 2,500 deaths in the local community. *Process Hazards Analysis*
1988	Piper Alpha, North Sea	Release of hydrocarbons led to an explosion and destruction of the offshore platform. 165 deaths. *Hot Work (LockOut)*
1989	Pasadena, TX	Release of ethylene/propylene led to a massive explosion. 23 deaths and about 130 injuries. *Mechanical Integrity, Training, Hot Work, Contractors*
1990	Channelview, TX	Explosion of storage tank. 21 deaths *Mechanical Integrity, Process Hazards Analysis*

All the elements of PSM are inter-connected. For example, if it is decided to work on the Emergency Action Plan, the following sequence of actions may occur.

1. The writing of the *Emergency Response Plan* (element 12 of Table 1-3) requires a knowledge of which hazards have to be addressed.

2. Consequently, a *Process Hazards Analysis* (element 3) is required to identify the hazards.

3. In order to be able to carry out the hazards analysis, information from sources such as P&ID's and MSDS is needed. Much of this information is part of *Process Safety Information* (element 2.)

4. Once the Emergency Response Plan has been developed, it will be necessary to *Train* everyone in its use (element 5.)

5. All of the above activities require *Employee Participation* (element 1.)

6. The Emergency Response Plan has to be *Audited* on a regular basis (element 13.)

It can be seen that this simple example involves interaction of six of the elements of PSM. Furthermore, it is likely that, during the training step, the people who are being trained will come up with ideas that will improve the quality of the emergency response plan. After going through the *Management of Change* step (element 10), these ideas can be used to upgrade the emergency manual.

When considered in isolation, many of the elements appear to be the "most important". For example, *Employee Participation* is the "most important" because, if the employees do not participate, the process safety program will not function properly. But *Management Of Change* could be considered the "most important" because the root cause of all incidents is uncontrolled change. On the other hand, all of the elements require a solid base of up-to-date, comprehensive information. Therefore *Process Safety Information* is the "most important." The real point, of course, is that they are all important and necessary, and they all rely on one another to be effective.

Another way of looking at the elements of process safety is to divide them that are concerned with fact-finding and those to do with management systems (analysis and action.) The first group evaluates progress toward goals, the second group is concerned with achieving those goals. Included in the first group are:

- Process Hazards Analysis
- Audits
- Incident Investigation
- Prestartup Safety Review
- Management Of Change
- Trade Secrets

Included in the second group are:

- Employee Participation
- Process Safety Information
- Operating Procedures
- Training
- Emergency Planning & Response
- Hot Work Permits
- Mechanical Integrity
- Contractors

MOST FREQUENTLY CITED PSM ELEMENTS

The Chemical Process Safety Report for June 1996 listed the number of OSHA citations for each of the elements for the period April 1, 1995 to April 5, 1996 for all the OSHA regions. The results are summarized in Table 1-5. The elements are sorted here according to the frequency with which they were cited.

Table 1-5

Most Frequently Cited Elements Of PSM

Element	Number of Citations	%
Mechanical Integrity	114	15
Process Hazard Analysis	111	15
Process Safety Information	108	15
Operating Procedures	107	14
Contractors	67	9
Employee Participation	63	8
Training	60	8
Management Of Change	48	6
Incident Investigation	23	3
Emergency Planning And Response	19	3
Pre-Startup Safety Review	10	1
Compliance Audits	8	1
Hot Work Permit	6	1
Total	**744**	**100**

Other surveys like this differ in detail, the general conclusion is that the top four items in Table 1-5 are the most frequently cited elements.

ORGANIZATION OF A PSM PROGRAM

The organization of a process safety program is discussed in detail in Chapter 17. One general approach that can be followed by companies in the United States is to make it part of the Risk Management Program that has to developed for the EPA (see Chapter 2 for details on this regulation.). Since the RMP is broader in scope than OSHA's PSM regulation, and has more requirements, the OSHA standard can be incorporated within it, using the following three-step approach.

1. Identify the hazards associated with the process, and determine their consequences.
2. Implement a management system to control the hazards (and their consequences).
3. Prepare a communications plan with the public so that any emergencies are handled properly.

The second of these three steps — the implementation of process safety — is very similar to the OSHA Process Safety Management program. Therefore, it is suggested that the overall process risk management program be developed according to the RMP rule. The PSM program will then be a major part of it. Using the PSM elements listed in the previous section, the overall organization, using this approach, is shown in Figure 1-2, which contains most of the OSHA PSM elements. (Many of the elements, such as Employee Participation, really belong in all three columns, but Figure 1-2 does show where they are the most important.)

Figure 1-2

RMP/PSM Management System

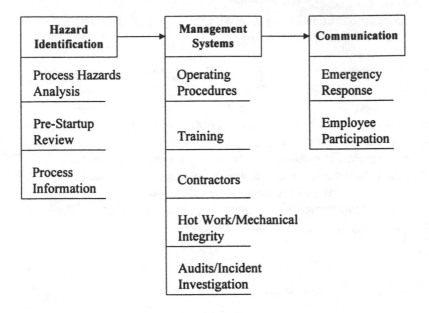

LEGAL PROTOCOL

This book concentrates on technical topics to do with process safety. It does not offer legal advice. However, it is appropriate to outline some of the legal protocol issues that all employees should be aware of, particularly with regard to documentation. If there is a serious accident, then the PSM documentation will be brought into the formal investigation process (unless protected by client/attorney privilege.) Those requiring the documentation may be representatives of an agency, such as OSHA, or the counsel representing plaintiffs in litigation. Therefore, the following points should be always be considered.

1. Documentation should confine itself to facts and sound engineering judgment. It should not contain any speculation, particularly to do with major accident scenarios.

2. Draft documents should be disposed of as soon as they are no longer relevant or correct. All drafts should be destroyed prior to the issue of a final report or document. Not only does this practice reduce the potential for problems in an investigation, it also reduces the chance of internal confusion. The control of draft documentation is difficult to enforce. It is likely that each person involved with the program will have his or her own copies of documents such as P&ID's and operating procedures. They will also probably have made notes in their diaries and files. If there is an accident, those notes becomes part of the official record. Yet people are reluctant to throw such documents away because "you never know when they might come in handy."

3. All recommendations and action items must be followed through to closure in a timely manner, and the closure should be properly documented.

4. Documentation should avoid the use of emotional or potentially inflammatory language. Generally, words such as 'catastrophe' and 'tragic' should be avoided. It should also avoid judgmental words like "serious" or "unacceptable." However, there is nothing wrong with using factual words or phrases such as 'fire' or 'rapid corrosion.'

5. When selecting a recommendation, it is important to show how it was chosen, particularly if it happened to be the cheapest among a variety of solutions.

6. There should never be any indication that safety was traded off against cost, nor that a safety recommendation was not implemented because it was too expensive.

7. The documentation should not imply that levels of "acceptable risk" have been determined (although the use of risk matrices and other risk calibration devices implicitly incorporates concepts of acceptable risk, as discussed in Chapter 16.)

    In addition to documentation issues, other legal items to watch for include the following.

8. There must never be any attempt to offer gifts to a regulatory official because the gift could be construed as being a bribe. Even offering a cup of coffee to an inspector should be thought about in this context.

9. All personnel should be taught how to answer audit questions, i.e. how to be open and honest, but not to volunteer any information that has not been asked for.

Additional guidance concerning working with regulatory agencies is provided in the next Chapter.

CONCLUSIONS

Process Safety Management is an on-going management process; it is not just to do with meeting an OSHA regulation. The application of PSM principles should help improve all aspects of a plant's operation — including safety, environmental and economic performance.

Process Safety Management never finishes; no matter how good a plant's safety record may be, there are always ways in which it can be improved. This is done by defining the proper control targets, and then making sure that the plant does not deviate outside them.

Chapter 2

REGULATIONS, STANDARDS & TRADE SECRETS

INTRODUCTION

 This Chapter provides an overview of process safety management regulations in the United States. Because this is a technical book, the information provided here cannot be used for legal interpretations of the regulations. Legal advice must be sought from a qualified attorney. Furthermore, most regulations contain material which is not generally of interest to technical personnel. For example, there is generally a preamble that defines the statutes that provide the authority for the creation of the regulation. In order to save space, this type of material has been excluded here. Irrelevant or extraneous information (such as target dates that have already expired) has also been omitted. Consequently, the material provided in this chapter can only be used as the basis of technical discussion and analysis. Any formal regulatory interpretation should be based on an official copy of the appropriate rule or regulation.

In the past, regulations were obtained in printed form from government and state printing offices. However, a quicker and more economical way of obtaining them is through the Internet. The material on the Net is up to date, it can be obtained right away and it is free. Both OSHA and the EPA have web sites at www.osha.gov and www.epa.gov respectively. It is simple to find the regulation of interest, and then either download it to a hard drive, or print it out.

Although regulations are often treated as being an unnecessary burden, they do help overcome procrastination, particularly with respect to low probability hazards. If a plant has been running safely for many years, it is tempting to defer the rectification of low probability hazards on the grounds that they have never been a problem in the past. Expenditures to correct these hazards do not lead to an immediate return on investment, all that they do is make an already low probability number even lower. The effect of a regulation is to force action to be taken in these cases.

For the same reason, some companies are reluctant to commission hazard analyses and other studies because, once a hazardous situation has been identified, they are obliged, both ethically and legally, to take corrective action. Once more, the existence of a regulation overcomes these reasons for not doing anything; the work has to be done, like it or not.

UNITED STATES FEDERAL REGULATIONS

A brief overview of how laws and regulations in the United States are written, interpreted and enforced is provided here.

THE REGULATORY PROCESS

In the United States, the federal regulatory process starts when Congress (both houses) proposes a law or statute. Generally, each house develops its own bill. These are then sent to committee, where a compromise bill is agreed upon. This then goes to the President, who signs it (unless he chooses to use his power of veto.) With regards to process safety, the statute that provides the basis for both OSHA's Process Safety Management (PSM) program and the EPA Risk Management Program (RMP) is the Amendments to the Clean Air Act, P.L. 101-549, signed into law by President Bush on November 15, 1990. This statute contains seven separate titles covering a wide range of regulatory programs. The risk management provisions are to be found under Title III (Hazardous Air Pollutants), part c (Accidental Chemical Releases.) They are sometimes referred to as a Chemical Accident Release Prevention (CARP) programs. Statutes specify which agencies are to interpret and enforce their provisions. The Clean Air Act statute required OSHA to implement its requirements with regard to worker safety, and the EPA for public safety.

Once a statute has been passed, the first responsibility of the affected agencies is to develop specific regulations from it. It is the regulation, not the law itself, that individuals and companies are expected to follow. When an agency has finished writing a regulation, it is listed in the <u>Code of Federal Regulations</u> (CFR), and indexed in the Federal Register (a daily publication of all federal government communications.) The public and other interested parties are then invited to comment on this draft regulation. This is usually done by mail and at public hearings. The hearings are held in Washington D.C. and in other cities where the proposed regulation is likely to have a major impact. One of the OSHA PSM hearings was held in Houston, Texas, because of the large number of process plants in that area.

At the conclusion of the hearings, all the comments are placed in a public docket. The reviewers consider these comments during the preparation of the final regulation, which is eventually published in the Federal Register. There is then a period of three to six months during which the first draft of the final regulation can be challenged in court. (This happened when OSHA introduced the PSM standard. The United Steelworkers of America and other labor unions challenged those parts of the regulation that covered contract workers. Also, some changes were made based on comments from organizations such as the Chemical Manufacturers Association and the Dow Chemical Company.)

Following the implementation of a standard, the agency can continue to make modifications and changes as circumstances dictate. Also, if a person or organization disagrees with some part of the regulation, they can challenge it in court on the grounds that it does not meet the intent of the original Congressional statute. OSHA has made some minor changes to the PSM standard since it was first issued in May, 1992.

GENERAL DUTY CLAUSES

Both OSHA and the EPA statutes contain General Duty Clauses that can be used to cover situations not explicitly identified by the regulations, but which nevertheless, in the judgment of the agency, fall within its purview. With regard to PSM and RMP, both of these clauses have been in effect from the moment the legislation was enacted, i.e. 1990. However, the General Duty Clause can only be applied if the agency can demonstrate that (a) the citation involved a "recognized hazard", and (b) that the hazard could cause death or serious physical harm.

One application of the General Duty clause with respect to process safety occurred following a serious accident in April, 1995, at a plant in Lodi, New Jersey. None of the chemicals involved in the accident were on the OSHA list of highly hazardous materials. OSHA's citations, therefore, were based on the General Duty clause. (Since then, the chemicals involved in that accident have been incorporated into the standard.)

THE TENTH AMENDMENT TO THE UNITED STATES CONSTITUTION

When discussing regulations in the United States, it is important to consider the implications of the Tenth Amendment to the Constitution, which reads as follows:

The powers not delegated to the United States by the Constitution, nor prohibited by it to the states, are reserved to the states respectively, or to the people.

This amendment is one of the foundations of the federal form of government. With respect to process safety, it means that any state has the right to promulgate its own standards as long as those standards are at least as stringent as the federal regulation. As long as this condition is met, the state regulation takes precedence over the federal regulation. With regard to OSHA, this is explained in the OSHA Instruction CPL 2-2.45A CH-1, dated September 13, 1994, which reads:

If the State adopts an alternative to Federal enforcement inspection procedures, the State's plan must identify and provide a rationale for all substantial differences from Federal procedures in order for OSHA to judge whether a different State procedure is as effective as the comparable procedure.

California is one state that chose to develop its own version of 29 CFR 1910.119 (in addition to its own separate process safety regulation known as the RMPP.) Their program, which is administered by CAL OSHA, was challenged by federal OSHA as being less rigorous than the federal standard. Implementation of the California standards was then blocked until they had been brought up to the level of the federal regulation.

States can also enact their own regulations; these may overlap with related federal regulations. With respect to process safety, three states — California, New Jersey and Delaware — have their own regulations.

Whenever the federal government requests that a state enforce a federal standard, federal funding should also be provided. At the time of writing (1997) the constitutionality of one federal law (the Brady bill) has been challenged on the grounds that states cannot be expected to enforce it given that the federal government has not provided any funds with which to do so.

THE OCCUPATIONAL SAFETY & HEALTH ADMINISTRATION (OSHA)

OSHA was formed in the year 1971 to provide standards for safety and health in the workplace (OSHA 2056, 1992.) Its jurisdiction covers the United States, Puerto Rico and federally administered territories. OSHA's mandate is restricted to the workplace. Although many of its actions will have impact "over the fence," this is essentially a secondary issue. Public safety and health is the responsibility of the Environmental Protection Agency.

It is estimated that there are about 6,000 injury deaths, 50,000 illness deaths and 7 million non-fatal injuries each year in industry within the United States. OSHA's goal is to reduce the number of these accidents by taking the following actions:

- Encourage employers and employees to reduce workplace hazards and to implement new or improve existing safety and health programs.

- Provide for research in occupational safety and health so as to develop innovative ways of dealing with occupational safety and health problems.

- Establish "separate but dependent responsibilities and rights" for employers and employees for the achievement of better safety and health conditions.

- Maintain a reporting and record-keeping system to monitor job-related injuries and illnesses.

- Establish training programs to increase the number and competence of occupational safety and health personnel.

- Develop mandatory job safety and health standards and enforce them effectively.

- Provide for the development, analysis, evaluation and approval of state occupational safety and health programs.

The agency has approximately 800 federal and 1,000 state compliance officers. (This is for all of the regulations that they administer — of which PSM is just a small part.) These are not large numbers given the scope of their responsibilities. Therefore, with respect to the process safety standard, it is unlikely that a company will be inspected unless there has been an accident or unless someone complains of an unsafe condition.

OSHA STANDARDS

A brief overview of the way in which OSHA standards are structured is provided here. The process safety management standard — 29 CFR 1910.119 — is used as an example. The system of headings and sub-headings that OSHA uses is as follows:

Level 1 — (a)
Level 2 — (1)
Level 3 — (i)
Level 4 — (A)
Level 5 — {1}

PART 29

Regulations in the Federal Register are located in "Parts." Almost all OSHA regulations are in Part 29, hence they usually start with the term 29 CFR, meaning Part 29 of the Code of Federal Regulations. Each part is then subdivided by number. Table 2-1 shows how Part 29 is divided. Process safety falls under 1910: Occupational Safety and Health Standards. Therefore, the regulations in this area have the prefix 29 CFR 1910. (This line has been shaded.)

Table 2-1

Parts of 29 CFR

PART 70	Production or Disclosure of Information or Materials
PART 70a	Protection of Individual Privacy in Records
PART 1900	Reserved
PART 1901	Procedures for State Agreements
PART 1902	State Plans for the Development and Enforcement of State Standards
PART 1903	Inspections, Citations, and Proposed Penalties
PART 1904	Recording and Reporting Occupational Injuries and Illness
PART 1905	Rules of Practice
PART 1906	Administration Witness and Documentation in Private Litigation
PART 1908	Consultation Agreements
PART 1910	Occupational Safety and Health Standards
PART 1911	Rules of Procedure for Promulgating, Modifying or Revoking OSHA Standards
PART 1912	Advisory Committees on Standards
PART 1912a	National Advisory Committee on OSHA

PART 1913	Rules Concerning OSHA Access to Employee Medical Records
PART 1915	Occupational. Safety and Health Standards for Shipyard Employment
PART 1917	Marine Terminal
PART 1918	Safety and Health Regulations for Longshoring
PART 1919	Gear Certification
PART 1920	Procedure for Variations under Longshoremen's Act
PART 1921	Rules of Practice in Enforcement under Section 41 of Longshoremen's Act
PART 1922	Investigative. Hearings under Section 41 of Longshoremen's Act
PART 1924	Safety Standards Applicable to Workshops and Rehab. Facilities
PART 1925	Safety and Health Standards for Federal Service Contracts
PART 1926	Safety and Health Regulations for Construction
PART 1927	Reserved
PART 1928	Occupational. Safety and Health Standards for Agriculture
PART 1949	Office of Training and Education, OSHA
PART 1950	Development and Planning Grants for Occupational Safety and Health
PART 1951	Grants for Implementing Approved State Plans
PART 1952	Approved State Plans for Enforcement of State Standards
PART 1953	Changes to State Plans for Development and Enforcement
PART 1954	Procedures for the Evaluation and Monitoring of Approved State Plans
PART 1955	Procedures for Withdrawal of Approval of State Plans
PART 1956	Plans for State and Local Government Employees without Approved Plans
PART 1960	Basic Program Elements for Federal Employees OSHA
PART 1975	Coverage of Employees under the Williams-Steiger OSHA 1970
PART 1977	Discrimination against Employees under OSHA Act of 1970
PART 1978	Rules for Implementing Section 405 of the STAA of 1982

| PART 1990 | Identification, Classification, and Regulation of Carcinogens |
| PART 2200 | Rules of Procedure |

SUBPARTS

Each part of the regulation is divided into Subparts, which are identified by letters. The Subparts for 29 CFR 1910 are shown in Table 2-2. Process safety is 29 CFR 1910 Subpart H — Hazardous Materials.

Table 2-2

Subparts of 29 CFR Part 1910

1910 Subpart A - General (1910.1 to 1910.8)
1910 Subpart B - Adoption and Extension of Established Federal Standards (1910.11 to 1910.19)
1910 Subpart C - General Safety and Health Provisions (1910.20 to 1910.20 App B)
1910 Subpart D - Walking-Working Surfaces (1910.21 to 1910.32)
1910 Subpart E - Means of Egress (1910.35 to 1910.40)
1910 Subpart F - Powered Platforms, Manlifts, and Vehicle-Mounted Work Platforms (1910.66 to 1910.70)
1910 Subpart G - Occupational Health and Environmental Control (1910.94 to 1910.100)
1910 Subpart H - Hazardous Materials (1910.101 to 1910.120 App E)
1910 Subpart I - Personal Protective Equipment (1910.132 to 1910.140)
1910 Subpart J - General Environmental Controls (1910.141 to 1910.150)
1910 Subpart K - Medical and First Aid (1910.151 to 1910.153)
1910 Subpart L - Fire Protection - Other Fire Protection Systems (1910.155 to 1910.165)
1910 Subpart M - Compressed Gas and Compressed Air Equipment (1910.166 to 1910.171)
1910 Subpart N - Materials Handling and Storage (1910.176 to 1910.190)

1910 Subpart O - Machinery and Machine Guarding (1910.211 to
          1910.222)
1910 Subpart P - Hand and Portable Powered Tools and Other Hand-
          Held Equipment. (1910.241 to 1910.247)
1910 Subpart Q - Welding, Cutting, and Brazing (1910.251 to
          1910.257)
1910 Subpart R - Special Industries (1910.261 to 1910.275)
1910 Subpart S - Electrical - Definitions (1910.301 to 1910.399)
1910 Subpart T - Commercial Diving Operations - Recordkeeping
          (1910.401 to 1910.441)
1910 Subpart U - [Reserved] (1910 Subpart U)
1910 Subpart V - [Reserved] (1910 Subpart V)
1910 Subpart W - [Reserved] (1910 Subpart W)
1910 Subpart X - [Reserved] (1910 Subpart X)
1910 Subpart Y - [Reserved] (1910 Subpart Y)
1910 Subpart Z - Toxic and Hazardous Substances (1910.1000 to
          1910.1500)

SECTIONS OF SUBPARTS

Subparts are divided using numbers. Subpart H is divided into the
range 101 to 120. The title for each of these is shown in Table 2-3.
Process safety is 119. Therefore, the full title of the OSHA PSM
standard is 29 CFR Subpart H 1910.119, often abbreviated to 29 CFR
1910.119, or even just 1910.119.

Table 2-3

29 CFR Part 1910

1910 Subpart H - Authority for 1910 Subpart H
1910.101 - Compressed gases (general requirements).
1910.102 - Acetylene.
1910.103 - Hydrogen.
1910.104 - Oxygen.
1910.105 - Nitrous oxide.
1910.106 - Flammable and combustible liquids.

1910.107 - Spray finishing using flammable and combustible materials.
1910.108 - Dip tanks containing flammable or combustible liquids.
1910.109 - Explosives and blasting agents.
1910.110 - Storage and handling of liquefied petroleum gases.
1910.111 - Storage and handling of anhydrous ammonia.
1910.112 - [Reserved]
1910.113 - [Reserved]
1910.114 - Effective dates.
1910.115 - Sources of standards.
1910.116 - Standards organizations.
1910.119 - Process safety management of highly hazardous chemicals.
1910.119 App A - List of Highly Hazardous Chemicals, Toxics and Reactives (Mandatory).
1910.119 App B - Block Flow Diagram and Simplified Process Flow Diagram (Nonmandatory).
1910.119 App C - Compliance Guidelines and Recommendations for Process Safety Management (Nonmandatory).
1910.119 App D - Sources of Further Information (Nonmandatory).
1910.120 - Hazardous waste operations and emergency response.
1910.120 App A - Personal protective equipment test methods.
1910.120 App B - General description and discussion of the levels of protection and protective gear.
1910.120 App C - Compliance guidelines.
1910.120 App D - References.
1910.120 App E - Training Curriculum Guidelines - (Non-mandatory)

29 CFR 1910.119

The rules for determining which facilities are covered by 29 CFR 1910.119 are shown in Table 2-4 (some of the non-technical details have been omitted.)

Table 2-4

Covered Processes

(I)	A process which involves a chemical at or above the specified threshold quantities . . .;
(ii)	A process which involves a flammable liquid or gas (as defined in 1910.1200(c) of this part) on site in one location, in a quantity of 10,000 pounds (4535.9 kg) or more except for:
(A)	Hydrocarbon fuels used solely for workplace consumption as a fuel (e.g., propane used for comfort heating, gasoline for vehicle refueling), if such fuels are not a part of a process containing another highly hazardous chemical covered by this standard;
(B)	Flammable liquids stored in atmospheric tanks or transferred which are kept below their normal boiling point without benefit of chilling or refrigeration
(2)	This section does not apply to:
(i)	Retail facilities;
(ii)	Oil or gas well drilling or servicing operations;
(iii)	Normally unoccupied remote facilities.

OSHA has prepared a list of covered chemicals (provided in Appendix A of the regulation.) If a company has one or more of these chemicals on site in quantities greater than the amount specified, then it is required to develop a Process Safety Management program. Even if a chemical is not on this list, an accident involving it can still lead to a PSM citation under the General Duty clause, as discussed earlier.

It is not always clear when a process is covered, and when it is not. Examples of potential ambiguity include the following.

- One section of a facility handles a hazardous chemical. Adjacent to it is another unit that does not handle hazardous chemicals, and that is not linked to the first unit by piping or any process equipment. However, the two are physically close. If there were to be an explosion on the non-covered unit, flying debris may puncture a tank in the neighboring plant, leading to the release of the hazardous chemical. This then raises the question as to whether process safety activities (such as hazards analyses) should be performed on the non-covered process.

- If a unit that does not handle hazardous chemicals is adjacent to one that does, the employees on the non-covered process may need to be trained in emergency response techniques should their neighbor have a release.

- Some plants are mostly unmanned; there is no full-time presence on the unit. With the advent of remotely operated DCS (distributed control systems,) this way of operating is now being used for quite sophisticated plants. This means that the operators may be hundreds of miles from the facility that they are controlling, thus posing a special challenge for those who are writing the Operating Procedures.

- Sometimes a plant may have a total inventory of the hazardous chemical that exceeds the threshold level, but that inventory is distributed among small vessels, none of which is above the threshold individually.

Each of these situations raises the question as to what processes are covered, and which ones are not. These question require legal clarification and/or advice from the appropriate regulatory agency. If a company wishes to establish that a process is not covered, even though there are other covered processes in the area, this decision should be backed up with a Process Hazards Analysis. If this analysis reasonably demonstrates that the non-covered process cannot be affected by the other plants, then this can be used in the event of a citation.

Another problem concerning the assignment of responsibility concerns the situation where a company owns equipment, but the equipment is located at another company's facility, and is operated and maintained by the people at the site.

THE ENVIRONMENTAL PROTECTION AGENCY (EPA)

The U.S. Environmental Protection Agency is charged by Congress to protect the Nation's land, air, and water systems. The EPA works in partnership with state, county, municipal, and tribal governments to carry out its mission. As with OSHA, State and local standards may implement their own rules as long as they exceed federal standards, but they cannot be less stringent.

The EPA administers 11 comprehensive environmental protection laws:

- Clean Air Act
- Clean Water Act
- Safe Drinking Water Act
- Comprehensive Environmental Response, Compensation, and Liability Act ("Superfund")
- Resource Conservation and Recovery Act
- Federal Insecticide, Fungicide, and Rodenticide Act
- Toxic Substances Control Act
- Uranium Mill Tailings Radiation Control Act
- Lead Contamination Control Act
- Ocean Dumping Ban Act
- National Environmental Education Act.

The Clean Air Act is the one that covers process safety. The rule is referred to as the Risk Management Program, and is administered by the Chemical Emergency Preparedness And Prevention Office (CEPPO), the EPA office that administers the Risk Management Plan (RMP.) Table 2-5 contains some material provided by the EPA about CEPPO.

Table 2-5

CEPPO

Historical Background

Public awareness of the potential danger from accidental releases of hazardous substances has increased over the years as serious chemical accidents have occurred around the world. Public concern intensified following the 1984 release of methyl isocyanate in Bhopal, India, which killed more than 2,000 people.

A subsequent chemical release in Institute, West Virginia, sent more than 100 people to the hospital and made Americans aware that such incidents can and do happen in the United States.

In response to this public concern and the hazards that exist, EPA created its Chemical Emergency Preparedness program (CEPP) in 1985: a voluntary program to encourage state and local authorities to identify hazards in their areas and to plan for potential chemical emergencies. This local planning complemented emergency response planning carried out at the national and regional levels by the National Response Team and Regional Response Teams.

The following year, Congress enacted many of the elements of CEPP in the Emergency Planning and Community Right-to-Know Act of 1986, also known as Title III of the Superfund Amendments and Reauthorization Act of 1986 (SARA). SARA Title III requires states to establish State Emergency Response Commissions (SERCs) and Local Emergency Planning Committees (LEPCs) to develop emergency response plans for each community. SARA Title III also requires facilities to make information available to the public on the hazardous chemicals they have on site. Title III's reporting requirements foster a valuable dialogue between industry and local communities on hazards to help citizens become more informed about the presence of hazardous chemicals that might affect public health and the environment.

EPA recognized that accident prevention, preparedness, and response are not discrete processes, but form a safety continuum. Therefore, in 1986, EPA established its Chemical Accident Prevention Program, integrating it with the Chemical Emergency Preparedness Program.

Under SARA section 305(b), EPA was required to conduct a review of emergency systems to monitor, detect, and prevent chemical accidents at facilities across the country. The final report to Congress, Review of Emergency Systems (EPA, 1988), concluded that the prevention of accidental releases requires an integrated approach that considers technologies, operations, and management practices, and it emphasized the importance of management commitment to safety.

The first prevention initiative was to begin collecting information on chemical accidents. Then EPA began working with other stakeholder groups to increase knowledge of prevention practices and encourage industry to improve safety at facilities.

Under the Chemical Accident Prevention Program, EPA developed the Accidental Release Information Program (ARIP) to collect data on the causes of accidents and the steps facilities take to prevent recurrences. EPA also developed its Chemical Safety Audit Program to gather and disseminate information on practices at facilities to mitigate and prevent chemical accidents. Another significant component of EPA's Chemical Accident Prevention Program involves outreach to small and medium-sized enterprises, which are generally less aware of risks than larger facilities.

All these efforts are based on the premise that while industry bears the primary responsibility for preventing and mitigating chemical accidents, many other groups also have a role to play. Workers, trade associations, environmental groups, professional organizations, public interest groups, the insurance and financial community, researchers and academia, the medical profession, and governments at all levels can help facilities that use hazardous chemicals identify their hazards and find safer ways to operate. A number of stakeholder groups have now developed programs and guidance to assist facilities in the management of chemical hazards. Many of these safety measures can make businesses more efficient and productive.

CEPPO Mission

EPA's CEPPO provides leadership, advocacy, and assistance to:

Prevent and prepare for chemical emergencies;
Respond to environmental crises; and
Inform the public about chemical hazards in their community.

To protect human health and the environment CEPPO develops, implements, and coordinates regulatory and non-regulatory programs.

THE EPA RISK MANAGEMENT PROGRAM — 40 CFR 68

Although the OSHA and EPA standards are purposely very similar to one another from a technical point of view, there are administrative differences, particularly with regard to terminology. For example, EPA uses the phrase "Owner or Operator" rather than "Employer," which is the term used by OSHA. Also, the RMP is referred to as a "rule," whereas PSM is a "regulation" or "standard."

TIERING/PROGRAM LEVELS

EPA recognizes that many companies handle just small amounts of chemicals, and that they do not have the resources to conduct full-scale risk analyses. Moreover, processes like these are often quite generic, and each plant is very similar to others of the same type. Therefore, EPA set up three Tiers or Program Levels. Program Level 1 is the simplest; Program Level 2 is somewhat more complex. Level 3 is the most stringent and is the one that applies to major process facilities. It covers nine SIC codes (the meaning of these codes is given later in this chapter): 2611, 2812, 2819, 2821, 2869, 2873, 2879 and 2911.

OSHA does not use a tiering approach with regard to PSM; if a company is covered by the standard, it must develop a program that leads to a safe operation — whatever that takes.

COVERED CHEMICALS

The RMP lists 77 toxic and 63 flammable substances. (The list is different from OSHA's.) Explosives and transportation are excluded. The following chemicals are included in the RMP standard, but not in 1910.119.

- Arsenous trichloride
- Boron trifluoride with methyl ether
- Chloroform
- Hydrazine
- Isopropyl chloroformate
- Methyl thiocyanate
- Tetranitromethane
- Titanium tetrachloride
- Toluene diisocyanate

The following chemicals have a lower concentration limit under the RMP standard than under the OSHA standard:

- Aqueous ammonia
- Formaldehyde
- Hydrogen chloride
- Hydrogen fluoride
- Nitric acid
- Oleum

FORMAL MANAGEMENT SYSTEM

One of the biggest differences between the RMP and PSM standards is that EPA requires that a formal, written program be prepared, and then placed in the public domain. OSHA does not require this. EPA needs a written plan because the RMP is very concerned with making sure that members of the public are fully informed as to what hazards are in their community, and what to do if there is an accident, and how the various emergency agencies can work together.

COMMUNITY PARTICIPATION

The OSHA standard does not discuss Community Participation because OSHA is concerned primarily with worker safety. However, this topic is of major concern within the EPA RMP.

In some communities, the process industries have a bad reputation, in spite of the fact that their actual safety record is good when compared to most other industries. Plants are often seen as places of mystery, where strange and dangerous substances are manufactured.

This perception that industry does not care can be exacerbated following an actual incident. For example, reporting on a fire in 1996 at a California refinery, the Chemical Process Safety Reporter (July 1996) contained these quotations:

> [the] plant manager at the . . . refinery, did not return phone calls

> [the spokesperson] did not return phone calls seeking further comment

> the three employees that [the refinery] considers to be responsible for the accident have been identified and the company has responded by suspending one employee for a week and firing the other two

To an outside observer, it appears as if neither the official spokesperson nor the manager considered the community's reaction as important. Furthermore, by disciplining the workers involved in the accident, there appears to be an attitude of blaming the operator rather than looking for root causes of the accident. In fact, it is likely that this management group did care very much about safety, but this is not the perception that was created with these actions.

WORST CASE RELEASE

One of the biggest difference between the EPA and the OSHA standards concerns offsite issues. EPA is more concerned about the long-range impact of an accident, particularly what will happen if a cloud of toxic gas gets "over the fence." With this in mind, one of the requirements of the first draft of the RMP rule was to prepare an absolute worst-case scenario, in which no credit is taken for any type of safety system (including passive systems such as earthen walls around a tank farm.) It was defined as a release over a 10 minute period of the largest quantity of a regulated substance on the facility resulting from a piping or vessel failure. If a facility handled both flammable and toxic materials, a worst case scenario needed to be prepared for each.

The worst-case requirement was controversial, and has since been reduced in scope. However, it did have the advantage of putting all facilities on the same terms so that different industries and hazards could be compared on equivalent bases. OSHA does not require a worst-case analysis, although some companies choose to carry them out anyway.

The modeling of vapor releases is complex, and requires the resources of skilled analysts (an overview of the topic is provided in Chapter 16.) Recognizing this, the EPA developed "look-up" tables where a facility would simply identify its particular situation and have a rough idea as to what the consequences of a release might be. These are particularly useful for Program (Tier) One and Two companies, which are typically quite similar to one another and which do not have the resources to conduct a full-scale risk analysis.

EMERGENCY PLAN

The EPA requires that a formal emergency plan be prepared, to include some of the off-site scenarios just touched on. OSHA requires that an emergency response system be in place, but not necessarily a formal plan involving the public.

FIVE-YEAR ACCIDENT HISTORY

The EPA rule requires that a five-year accident history be prepared for the facility. This report should include injuries or fatalities on or off the site during the five year period. However, it does not have to include information about near-miss situations.

INVESTIGATION BOARD

The Act required that a five-member Chemical Safety and Hazard Investigation Board be established. Members would be appointed based on their technical knowledge of safety, engineering and environmental issues. This Board was never created due to funding limitations.

STATE REGULATIONS

Four states — California, New Jersey, Delaware and Nevada — have their own risk management plans, and twenty-five have set up their own version of the OSHA standard.

CALIFORNIA — RISK MANAGEMENT PREVENTION PROGRAM

The California Risk Management and Prevention Program (RMPP) was the first major regulation of its type in the United States. Like the EPA RMP, the RMPP requires that a written report be submitted.

NEW JERSEY — TOXIC CATASTROPHE PREVENTION ACT

Companies in New Jersey are required to meet the Toxic Catastrophe Prevention Act (TCPA); it is administered by the New Jersey Department of Environmental Protection and Energy (DEPE.)

One of the key features of the New Jersey standard is the definition of Extremely Hazardous Substance (EHS.) (Other regulations simply list the covered chemicals without providing much explanation as to what process what used for including chemicals in the list.) Companies which handle, use, manufacture or store, or have the capability of generating an EHS within one hour, are included in the standard. A list of EHS's is provided, along with the Registration Quantities (RQ) for each. The Registered Quantity does not all have to be in one location. The agency specifically states: "the RQ which triggers the requirements of the TCPA rule is 'site' based, rather than 'facility' based. Therefore, two or more facilities located on the same site, each with less than the RQ of an EHS will nonetheless be subject to the requirements of the Act and the rules if the total quantities of the EHS exceeds the registered quantity."

Furthermore, all equipment that handles the EHS, even if some of that equipment is away from the site where the RQ of the EHS is located, is included in the rule. There is, however, some latitude if a hazard analysis shows that the concentration of the EHS does not exceed a certain level beyond the site boundary.

Another feature of the TCPA is that it is considerably more prescriptive than most PSM programs. The regulation is comparatively detailed.

DELAWARE/NEVADA

Delaware's Extremely Hazardous Substances Risk Management Act was passed in 1988. It covers 89 regulated substances. Nevada's Senate Bill No. 641 was passed in 1991.

INDUSTRY STANDARDS

Many companies follow standards provided by professional bodies such as the American Petroleum Institute (API), the Chemical Manufacturers Association (CMA), the Center for Chemical Process Safety (CCPS) and the National Fire Protection Association (NFPA.) Although the standards and guidance issued by these bodies is not law, they possess a high degree of credibility and authority when a company is planning its safety work. A list of Industry Associations is provided in <u>Chemical Processing</u> (February, 1997.)

AMERICAN PETROLEUM INSTITUTE — RP 750

The American Petroleum Institute (API), founded in 1919, is the U.S. petroleum industry's primary trade association. Its membership consists of a broad cross section of petroleum and allied industries in exploration, production, transportation, refining and marketing. The API's membership currently includes more than 300 companies.

API provides public policy development, advocacy, research and technical services to enhance the ability of the petroleum industry to fulfill its mission, which includes:

- Meeting the nation's energy needs, developing energy sources, and supplying high-quality products and services.
- Enhancing the environmental, health, and safety performance of the petroleum industry.
- Conducting research to advance petroleum technology, and developing industry equipment and performance standards.
- Advocating government decision-making to encourage efficient and economic oil and natural gas development, refining, transportation and use; promoting public understanding of the industry's value to society, and serving as a forum on issues affecting the petroleum industry.

API activities are led by committees composed of representatives from all sectors of the industry and from companies of all sizes. Through these committees, API determines policy and positions on issues affecting the industry; plans programs and activities; deals with problems and developments of industry concern, and provides a wide range of services to member companies. The API Process Safety standard, Recommended Practice (RP) 750, which was issued on January 1990, is very similar in content to the OSHA standard because it was developed from the ORC guidance (as discussed in Chapter 1.)

CHEMICAL MANUFACTURERS ASSOCIATION

The Chemical Manufacturers Association (CMA), established in 1872 as the Manufacturing Chemists Association, represents the interests of the chemical industry. One of its activities is to research ways to minimize risks to employees and to the environment from chemical plants. CMA activities include communicating with the government and the public in the areas of taxation, environmental and workplace safety regulations, and engineering and safety standards. CMA committees address such issues as energy, environmental management, health and safety, government relations, risk policy, trade, and communications. CMA programs and services include a library, industry-government forums, awards, 24-hour toll-free emergency advice on chemical transport accidents, CHEMTREC, speakers, conferences, and radio programs, a lending library of videotapes for emergency response training, and a chemical information and referral center.

In 1988 the Association launched its Responsible Care® program based on the concept developed by the Canadian Chemical Producers Association (Chemical Engineering, March 1997) to improve the industry's responsible management of chemicals. Participation in the initiative is a requirement of association membership. The Community Awareness and Response (CAER) program is a part of the Responsible Care initiative.

CENTER FOR CHEMICAL PROCESS SAFETY

The Center For Chemical Process Safety (CCPS) is a Directorate of the American Institute Of Chemical Engineers (AIChE.) The CCPS was established in 1985 in response to a number of catastrophic incidents that had occurred prior to that time. CCPS is sponsored by approximately 80 companies in the process industries. It aims to improve process safety by publishing books, organizing conferences, conducting classes and sponsoring research. The CCPS is probably best known through its very extensive list of books to do with process safety.

ISO 9000/ISO 14001

ISO standards are concerned with quality, not just safety. They are quite similar to process safety in they way that they are implemented. There are no externally mandated standards; instead, each company has to develop standards that are appropriate for its circumstances. In essence, a company which is implementing ISO 9000 does four things:

1. It writes down what it is going to do.
2. It trains everybody in the standards that have been set.
3. It implements an audit program.
4. It suggests means for improving the present operation.

ISO 14001 is similar to ISO 9000 except that it focuses on environmental compliance. It incorporates, but goes beyond, legal requirements on environmental issues. At the time of writing, the ISO 14001 standard had not been finalized.

OTHER STANDARDS

Many parts of the OSHA PSM regulation refer to other OSHA standards. For example, the Emergency Planning & Response element basically requires that 29 CFR 1910.38 — Emergency Plans And Fire Prevention — be properly implemented. Other related standards include 1910.147 (Lockout/Tagout) and 1910.165 (Employee Alarm Systems.) Even when it is not necessary to follow them, these standards can still provide good general guidance.

Other regulations from other agencies are sometimes part of a process safety program. For example, with regard to docking facilities for ships and barges, OSHA normally covers the dock area and the dock employees. The Coast Guard normally covers the ship and the ship employees. Truck and rail safety is covered by various Department of Transportation (DOT) regulations. Sometimes it can be difficult to determine who has jurisdiction over what.

SIC CLASSIFICATIONS

Industrial activities are divided into groups known as Standard Industrial Classifications, or SIC's. These codes are sometimes referred to in safety regulations, hence it is useful to have a general understanding of them. At the top level of the SIC system are Divisions. These are then sub-divided into Major Groups, as shown in Table 2-6. Those to do with the process industries have been shaded. (Further information on recent changes to the SIC Classifications can be obtained at www.ntis.gov/business/sic.htm and www.osha.gov/cgi-bin/sic/sicser5.)

Table 2-6

SIC Divisions And Major Groups

Division B: Mining

Major Group 10: Metal Mining
Major Group 12: Coal Mining
Major Group 13: Oil And Gas Extraction
Major Group 14: Mining And Quarrying Of Nonmetallic Minerals,
Except Fuels

Division C: Construction

Major Group 15: Building Construction General Contractors And
Operative Builders
Major Group 16: Heavy Construction Other Than Building
Construction Contractors
Major Group 17: Construction Special Trade Contractors

Division D: Manufacturing

Major Group 20: Food And Kindred Products
Major Group 21: Tobacco Products
Major Group 22: Textile Mill Products
Major Group 26: Paper And Allied Products
Major Group 27: Printing, Publishing, And Allied Industries
Major Group 28: Chemicals And Allied Products
Major Group 29: Petroleum Refining And Related Industries
Major Group 30: Rubber And Miscellaneous Plastics Products
Major Group 31: Leather And Leather Products
Major Group 32: Stone, Clay, Glass, And Concrete Products
Major Group 33: Primary Metal Industries
Major Group 34: Fabricated Metal Products, Except Machinery And
                Transportation Equipment
Major Group 35: Industrial And Commercial Machinery And
                Computer Equipment
Major Group 37: Transportation Equipment

Major Group 38: Measuring, Analyzing, And Controlling Instruments; Photographic, Medical And Optical Goods; Watches And Clocks

Major Group 39: Miscellaneous Manufacturing Industries

Division E: Transportation, Communications, Electric, Gas, And Sanitary Services

Major Group 40: Railroad Transportation
Major Group 41: Local And Suburban Transit And Interurban Highway Passenger Transportation
Major Group 42: Motor Freight Transportation And Warehousing
Major Group 43: United States Postal Service
Major Group 44: Water Transportation
Major Group 45: Transportation By Air
Major Group 46: Pipelines, Except Natural Gas
Major Group 47: Transportation Services
Major Group 48: Communications
Major Group 49: Electric, Gas, And Sanitary Services
Major Group 87: Engineering, Accounting, Research, Management, And Related Services

Most of the process industries are a part of Major Group 28. This includes companies producing basic chemicals, and those establishments that manufacture products by predominantly chemical processes. Industries in this major group manufacture three classes of products:

1. Basic chemicals, such as acids, alkalis, salts, and organic chemicals.

2. Chemical products to be used in further manufacture, such as synthetic fibers, plastics materials, dry colors, and pigments.

3. Finished chemical products to be used for ultimate consumption, such as drugs, cosmetics, and soaps; or to be used as materials or supplies in other industries, such as paints, fertilizers, and explosives.

Establishments primarily engaged in packaging, repackaging, and bottling of purchased chemical products, but not engaged in manufacturing chemicals and allied products, are classified in Wholesale or Retail Trade industries.

A summary of the industries within SIC Major Group 28 is provided in Table 2-7.

Table 2-7

Major Group 28

Industry Group 281: Industrial Inorganic Chemicals

2812 Alkalis And Chlorine
2813 Industrial Gases
2816 Inorganic Pigments
2819 Industrial Inorganic Chemicals, Not Elsewhere Classified

Industry Group 282: Plastics Materials And Synthetic Resins, Synthetic

2821 Plastics Materials, Synthetic Resins, And Nonvulcanizable Elastomers
2822 Synthetic Rubber (Vulcanizable Elastomers)
2823 Cellulosic Manmade Fibers
2824 Manmade Organic Fibers, Except Cellulosic

Industry Group 283: Drugs

2833 Medicinal Chemicals And Botanical Products
2834 Pharmaceutical Preparations
2835 In Vitro And In Vivo Diagnostic Substances
2836 Biological Products, Except Diagnostic Substances

Industry Group 284: Soap, Detergents, And Cleaning Preparations;

2841 Soap And Other Detergents, Except Specialty Cleaners
2842 Specialty Cleaning, Polishing, And Sanitation Preparations

2843 Surface Active Agents, Finishing Agents, Sulfonated Oils, And
2844 Perfumes, Cosmetics, And Other Toilet Preparations

Industry Group 285: Paints, Varnishes, Lacquers, Enamels, And Allied

2851 Paints, Varnishes, Lacquers, Enamels, And Allied Products

Industry Group 286: Industrial Organic Chemicals

2861 Gum And Wood Chemicals
2865 Cyclic Organic Crudes And Intermediates, And Organic Dyes
And
2869 Industrial Organic Chemicals, Not Elsewhere Classified

Industry Group 287: Agricultural Chemicals

2873 Nitrogenous Fertilizers
2874 Phosphatic Fertilizers
2875 Fertilizers, Mixing Only
2879 Pesticides And Agricultural Chemicals, Not Elsewhere Classified

Industry Group 281: If The Chemicals Are Inorganic And In Industry
Group 286 If They Are

Industry Group 289: Miscellaneous Chemical Products

2891 Adhesives And Sealants
2892 Explosives
2893 Printing Ink
2895 Carbon Black
2899 Chemicals And Chemical Preparations, Not Elsewhere Classified

WORKING WITH REGULATORY AGENCIES

This section provides some guidance regarding working with regulatory agencies. For any specific situation, advice should be sought from a qualified attorney. Additional discussion concerning legal protocol is provided on page 20.

VARIANCES

Should a company be unable to meet the requirements of an OSHA regulation, it can apply for a variance. The basis for this can be: "shortage of materials, equipment or professional or technical personnel." A variance can also be applied for if the company's existing practices are "at least as effective as that required by OSHA."

With regard to the PSM standard, companies will sometimes apply for a temporary variance. This is similar to the full variance except that the company acknowledges that they fully intend to comply, but they need more time. In such cases, "Employers must demonstrate to OSHA that they are taking all available steps to safeguard employees in the meantime . . . " This is particularly important for PSM because so many of the elements were fully effective as soon as the standard was enacted in 1992; there was no grace period. Companies frequently found that they did in fact need time to get many of the elements up to speed. Variances allowed them to do so.

Temporary variances will not normally be given for a period of more than one year, and they cannot be renewed more than twice. Finally, "the temporary variance will not be granted to an employer who simply cannot afford to pay for the necessary alterations, equipment, or personnel."

CITATIONS

If the OSHA inspector believes that a violation of the standard has taken place, he or she may choose to issue one or more citations. Citations are categorized as follows:

- *Willful.* Willful (and egregious) violations are the most serious form of violation. They are committed by an employer with intentional disregard of the standard. OSHA must generally demonstrate that the employer knew about the facts of the cited condition, and knew that the regulation required that action be taken.

- *Serious.* A serious violation is one where there is a substantial probability that death or serious injury could result from the cited condition, and the employer either knew about, or should have known, about this condition. Knowledge of the regulation is not required in this case.

- *Other-than-Serious.* This is a violation where the hazard does not present a substantial probability of death or serious physical injury. This level of citation is not likely to be used with regards to PSM because, according to OSHA: "Any violation of the PSM standard . . . is a condition which could result in death or serious physical harm to employees. Accordingly, violations of the PSM standard shall not normally be classified as 'other-than-serious'." (CPL 2-2.45A.)

- *Repeat.* A repeat violation is one that occurs more than once and that is noted on two separate inspections.

- *Failure-to-Abate.* This is similar to the Repeat citation, in that an employer has failed to correct a condition that was previously cited.

Associated with each citation is a proposed penalty. If the employer chooses to contest the alleged violations or proposed penalties, they first discuss it with the agency at a conference. If that does not lead to a solution, the case can be presented to an independent Occupational Safety and Health Review Commission (OSHRC.)

The OSHRC is independent of both OSHA and the Department of Labor. It employees Administrative Law Judges (ALJ's) to decide on disputed citations and penalties. There are also appeal levels above the individual ALJ.

Some companies choose to challenge citations, even when the proposed fine is small. There are two reasons for this. First, there is a chance that the citation will be rejected. Up to 80% of citations have been rejected on the grounds that it contained an error that invalidated them. The second reason for challenging a citation is that it is important not to build up a record of incidents because each subsequent incident will lead to increasingly heavy fines. If, for example, a company is cited for a deficiency and a $5,000 fine is proposed, they might decide to accept it on the grounds that it is not worth the cost of fighting. However, if there is a second citation, and the first one is on the record, then the second citation might go to $50,000.

However, not all managers agree with this approach to regulatory citations. Some feel that the time and effort spent battling the citations could be better spent on getting on with business and improving safety. Therefore, they choose settle with the agency as quickly as possible so as to minimize the distraction caused by the dispute.

ENFORCEMENT

Many environmental and safety regulations carry criminal liability provisions. This means that a person who breaks the law can be prosecuted under the criminal code. Furthermore, if anyone else approved of that act, or failed to stop it, knowing that a violation was taking place, then that person is also liable to criminal prosecution. In addition, a corporate officer can be held vicariously liable for the conduct of a subordinate employee. This also applies if that officer deliberately screens himself so as not to hear the facts. Although this legal situation is normally only in the background of most PSM programs, it is important that plant personnel know the seriousness of the regulations and the consequences of non-compliance.

Examination of the records of citations show that, in many cases, the size of the fine is small as compared to the overall economics of the companies affected. However, the simple fact of being cited can be very detrimental to the reputations of those involved. In the broader context of law enforcement, people who are basically law-abiding are deterred from doing something wrong not so much by the fear of punishment, but by the fear of getting caught. In the case of PSM, managers may be more concerned with the career implications of having their facility cited than they are with the magnitude of the fines themselves.

WORKING WITH OSHA

OSHA will not generally inspect a facility unless there has been an accident or someone has made a complaint concerning an unsafe existence. The following guidelines are used by OSHA to set priorities for their inspections.

1. If there is a fatality or an accident that results in the hospitalization of five or more workers, it must be reported to the nearest OSHA office within 48 hours.

2. If OSHA is reasonably certain that a condition or situation exists that could cause death or serious injury before normal enforcement procedures would prevent it.

3. If an employee believes that there is an imminent hazard, he or she may ask for an OSHA inspection. OSHA is not obliged to respond to such requests; it will use judgment as to which to follow up. This helps the agency avoid vindictive or trivial complaints.

4. OSHA may target specific industries which have been identified as having the potential for serious problems.

5. Follow up to check on progress from a previous citation.

OSHA INSPECTIONS

OSHA has its own protocol for inspections. It can be found in OSHA Instruction CPL 2-2.45A. It is based on the Program-Quality-Verification (PQV) inspection method, and is discussed in Chapter 15 of this book (Compliance Audits.)

Each company should establish working guidelines as to how to work with regulators from agencies such as OSHA and the EPA during an inspection. These guidelines will involve issues such as:

1. How to react to an unexpected visit from an inspector
2. How to present the PSM documentation
3. How to respond to citations

In all these cases, the guidance must be developed in conjunction with the company's own attorneys. Some general discussion points are provided below.

THE ENTRY PROCESS

Strictly speaking, an OSHA inspector cannot enter a facility to conduct an inspection without having a warrant. In practice, common sense and common courtesy dictate that, should an inspector show up unannounced and without a warrant, it is appropriate to allow them into the facility (once they have presented their official identification, which should include a number to call so that the facility knows that the visit is authentic.) As an OSHA DATA report points out, inspectors are only human. The report shows that those companies that deny entry receive nearly twice as many citations as those that do not. Furthermore, the average penalty is twice as large. One reason for this may simply be that those companies who deny entry initially may have more problems to hide than those which are more open. Nevertheless, it does suggest that cooperation with inspectors is effective.

Even when the inspector does have a warrant, it is possible to challenge the basis on which that warrant was obtained. However, these are non-technical decisions that should be handled by the company's attorneys.

It is illegal for anyone to notify an employer that an OSHA audit is to be carried out.

REASON FOR ENTRY

There are various reasons why OSHA or some other agency may decide to carry out an inspection.

1. *Fatality / Catastrophe.* This is usually the most serious. Such investigations, often referred to as FAT/CAT, are performed when a very serious accident has occurred. This means that there has

been a fatality, or five or more people have been hospitalized.

2. *Programmed Inspections.* These are planned at the beginning of the year by the OSHA regional offices. OSHA attempts to develop objective criteria for its choice of facilities to inspect so as to justify the warrant-application process.

3. *Complaints.* If a worker complains about unsafe conditions, the agency may follow up with an inspection. Although complaints are an important source of information about unsafe conditions, they can be used by workers merely to express general grievances against management. An analysis by OSHA DATA shows that 19% of all OSHA inspections are in response to an complaint. But the same analysis shows that a high proportion (31%) of these complaints do not result in a citation.

THE INSPECTION PROCESS

OSHA audits are typically divided into five parts:

1. Kick-Off Meeting
2. Plant Tour
3. Records Review
4. On-Site Inspection
5. Interviews

An OSHA inspection will open with a kick-off meeting. At that time, the inspector will provide the employer with a copy of the standard and a letter notifying the employer that the facility is covered by the standard. The inspector, or Compliance Safety and Health Officer (CSHO), must have received the appropriate OSHA training. During the kick-off meeting, the OSHA officer must explain the purpose of the visit, and which parts of the plant are to be inspected. During the visit, the OSHA inspector will take notes, collect information and take pictures. It is perfectly in order for a representative of the company to take a matching set of notes and pictures in the event that there should be a dispute with OSHA's findings. Also, employees have a right to have one of their representatives present during an inspection tour. If there is a union, it will select the person; otherwise the employees will have to find some other way of finding a representative. This person cannot be appointed by management.

The inspection will usually start with a plant tour, during which the inspector will obtain a general overview of the facility and its operations. During this tour, a management representative may accompany the inspector (unless he or she wishes to speak to an employee privately.) At the conclusion of the tour, the OSHA officer must inform the plant management of any violations found, along with a statement as to how much time is to be allowed for correction.

The inspector will ask to see documentation for each of the elements of the standard. He or she will probably also ask for copies of the OSHA 200 logs for the past three years. The inspector will want information from both the facility owner and from the management of any contractors that are on site. The inspector will use the findings from the preliminary walk-around and the documentation to assess which units will be evaluated for compliance.

Further information on the OSHA enforcement process is provided in the April 1997 edition of the Chemical Process Safety Report. This document provides a thorough review of the inspection process, the way in which citations are issued, the role of the Occupational Safety and Health Review Commission, types of violation and enforcement policies.

TRADE SECRETS

The OSHA standard contains fourteen elements. Of these, the topic of Trade Secrets is the only one that is non-technical. Basically, OSHA wants to ensure that employees have access to the information that they need to carry out their job safely, regardless of Trade Secret concerns. A Trade Secret is some knowledge that is not formally patented, but that nevertheless gives companies a competitive edge. The term is defined by OSHA elsewhere as follows:

A trade secret may consist of any formula, pattern, device or compilation of information which is used in one's business, and which gives him an opportunity to obtain an advantage over competitors who do not know or use it. It may be a formula for a chemical compound, a process of manufacturing, treating or preserving materials, a pattern for a machine or other device, or a list of customers. It differs from other secret information in a business (. . .) in that it is not simply information as to single or ephemeral events in the conduct of the business, as, for example, the amount or other terms of a secret bid for a contract or the salary of certain employees, or the security investments made or contemplated, or the date fixed for the announcement of a new policy or for bringing out a new model or the like. A trade secret is a process or device for continuous use in the operations of the business. Generally it relates to the production of goods, as, for example, a machine or formula for the production of an article. It may, however, relate to the sale of goods or to other operations in the business, such as a code for determining discounts, rebates or other concessions in a price list or catalogue, or a list of specialized customers, or a method of bookkeeping or other office management.

Secrecy

The subject matter of a trade secret must be secret. Matters of public knowledge or of general knowledge in an industry cannot be appropriated by one as his secret. Matters which are completely disclosed by the goods which one markets cannot be his secret. Substantially, a trade secret is known only in the particular business in which it is used. It is not requisite that only the proprietor of the business know it. He may, without losing his protection, communicate it to employees involved in its use. He may likewise communicate it to others pledged to secrecy. Others may also know of it independently, as, for example, when they have discovered the process or formula by independent invention and are keeping it secret. Nevertheless, a substantial element of secrecy must exist, so that, except by the use of improper means, there would be difficulty in acquiring the information. An exact definition of a trade secret is not possible. Some factors to be considered in determining whether given information is one's trade secret are: (1) The extent to which the information is known outside of his business; (2) the extent to which it is known by employees and others involved in his business; (3) the extent of measures taken by him to guard the secrecy of the information; (4) the value of the information to him and his competitors; (5) the amount of effort or money expended by him in developing the information; (6) the ease or difficulty with which the information could be properly acquired or duplicated by others.

Novelty and prior art

A trade secret may be a device or process which is patentable; but it need not be that. It may be a device or process which is clearly anticipated in the prior art or one which is merely a mechanical improvement that a good mechanic can make. Novelty and invention are not requisite for a trade secret as they are for patentability. These requirements are essential to patentability because a patent protects against unlicensed use of the patented device or process even by one who discovers it properly through independent research. The patent monopoly is a reward to the inventor. But such is not the case with a trade secret. Its protection is not based on a policy of rewarding or otherwise encouraging the development of secret processes or devices. The protection is merely against breach of faith and reprehensible means of learning another's secret. For this limited protection it is not appropriate to require also the kind of novelty and invention which is a requisite of patentability. The nature of the secret is, however, an important factor in determining the kind of relief that is appropriate against one who is subject to liability under the rule stated in this Section. Thus, if the secret consists of a device or process which is a novel invention, one who acquires the secret wrongfully is ordinarily enjoined from further use of it and is required to account for the profits derived from his past use. If, on the other hand, the secret consists of mechanical improvements that a good mechanic can make without resort to the secret, the wrongdoer's liability may be limited to damages, and an injunction against future use of the improvements made with the aid of the secret may be inappropriate.

The following situations may require that employees have access to material that may be protected by Trade Secret law.

- Development and compilation of Process Safety Information
- A Process Hazards Analysis investigation
- Development of Operating Procedures
- Incident Investigation

- Preparation of Emergency Plans
- Audits

The PSM regulation concerning Trade Secrets is provided below. The key words in terms of implementing process safety are in the final clause of paragraph (1).

---

(1) Employers shall make all information necessary to comply with the section available to those persons responsible for compiling the process safety information (required by paragraph (d) of this section), those assisting in the development of the process hazard analysis (required by paragraph (e) of this section), those responsible for developing the operating procedures (required by paragraph (f) of this section), and those involved in incident investigations (required by paragraph (m) of this section), emergency planning and response (paragraph (n) of this section) and compliance audits (paragraph (o) of this section) without regard to possible trade secret status of such information.

(2) Nothing in this paragraph shall preclude the employer from requiring the persons to whom the information is made available under paragraph (p)(1) of this section to enter into confidentiality agreements not to disclose the information as set forth in 29 CFR

(3) Subject to the rules and procedures set forth in 29 CFR 1910.1200(i)(1) through 1910.1200(i)(12), employees and their designated representatives shall have access to trade secret information contained within the process hazard analysis and other documents required to be developed by this standard.

Usually, a Trade Secret is used to protect some technique or operation that the company has developed to give itself a competitive advantage. Unlike new technology, ideas and techniques covered by Trade Secrets cannot be patented. OSHA recognizes that there is a potential conflict between the Employee Participation of the PSM standard and a company's right to protect its competitive advantages. OSHA insists that employees must have access to the information that they need to carry out their work safely. However, there are ways in which employers can protect their rights.

First, an employer can ask employees to sign a confidentiality agreement. In this agreement, employees (including contract workers) will agree to use the protected information for safety purposes only, and they agree not to divulge it to anyone outside the company. Second, a company need only reveal the information necessary for safety. For example, if the employees need to know the flash point for a chemical, but this does not mean that the company must also reveal the formula of that chemical. Third, companies can use code words to hide sensitive information. If they have a proprietary catalyst, for example, they may simply choose to identify as X-12, rather than revealing its correct chemical name.

Table 2-8 provides an example of a Trade Secrets secrecy form that an employee could be asked to sign.

Table 2-8

Representative Trade Secret Form

I, _____ (name, address and other personal information here)

agree to keep all information in confidence relating to the design of equipment, process flow schemes, and yields less than six months old that I might learn from the records and files of COMPANY, and not to transmit or discuss this information with persons or organizations that have not entered into a secrecy agreement with or release from COMPANY. This agreement is in exchange for being allowed to enter the files and records of COMPANY for information pertaining to §1910.119, Process Safety Management of Highly Hazardous Materials.

CONCLUSIONS

For many companies, regulations were needed to force them to implement a formal process safety program. In some cases, process safety work had been indefinitely deferred because there is not likely to be an immediate, short-term pay-out (although there will certainly be a long-term improvement in performance.) The regulations have been written in a non-prescriptive manner so that companies can do the right thing toward improving safety, productivity and environmental performance.

CONCLUSION

Chapter 3

EMPLOYEE PARTICIPATION

INTRODUCTION

 If a PSM program is to be successful, it must involve all the people working at the site: managers, hourly employees, contract workers and administrative staff. There are three reasons for this. The first is that everyone wants to feel involved in the company's activities, and to know that their opinions are being heard. This involvement is the foundation for their commitment to safety. Second, it is the people at the working level who know most about the plant operations, and are able to identify areas where improvements can be made. Third, Employee Participation provides a sanity check on ideas, projects and analyses. Anything new or unusual should be reviewed by the employees; they will immediately identify any common sense problems.

For these reasons, Employee Participation can be regarded as the foundation of all process safety activities. It is not a stand-alone activity; instead it should be woven into the fabric of all the elements. Its purpose is to help employees plan and implement each element of standard, thus making the overall program more effective. For example, the purpose of Process Hazards Analysis (PHA) is not just to identify hazards, but also to encourage a particular way of thinking among all employees. So, if an operator working by himself is about to open a valve, before doing so he might spend a few moments going through some of the PHA guidewords (like "reverse flow.") While doing this, he may identify a possible accident situation, and decide not to perform that action until it has been reviewed a little more thoroughly. When this happens, both the Employee Participation and Process Hazards Analysis elements are working well.

It can be seen, therefore, that Employee Participation is not just about employee *communication*, it is about *participation*. It is not sufficient just to make information available to operators, they must be involved in the creation and execution of all parts of the process safety program. For example, the OSHA standard requires that employees be consulted regarding the conduct and development of Process Hazards Analyses. Simply placing a copy of the PHA report in the control room does not satisfy this requirement. Instead, the PHA teams should always include operators (often on a rotating assignment), and all findings should be thoroughly explained to those who were not on the team. They should be encouraged to comment on what they have been told.

THE OSHA REGULATION

(1) Employers shall develop a written plan of action regarding the implementation of the employee participation required by this paragraph.

(2) Employers shall consult with employees and their representatives on the conduct and development of process hazards analyses and on the development of the other elements of process safety management in this standard.

(3) Employers shall provide to employees and their representatives access to process hazard analyses and to all other information required to be developed under this standard.

OSHA GUIDANCE

OSHA provides the following guidance with regard to Employee Participation (some of the preamble material has been removed.)

Employers are to consult with their employees and their representatives regarding the employers efforts in the development and implementation of the process safety management program elements and hazard assessments. [Employers must] train and educate their employees and to inform affected employees of the findings from incident investigations required by the process safety management program. Many employers, under their safety and health programs, have already established means and methods to keep employees and their representatives informed about relevant safety and health issues and employers may be able to adapt these practices and procedures to meet their obligations under this standard. Employers who have not implemented an occupational safety and health program may wish to form a safety and health committee of employees and management representatives to help the employer meet the obligations specified by this standard. These committees can become a significant ally in helping the employer to implement and maintain an effective process safety management program for all employees.

ANALYSIS OF THE REGULATION

(1) Written Plan Of Action

OSHA requires that the Employee Participation program be written down. This can be difficult to do well because Employee Participation is involved in so many areas of process safety. The plan of action should identify who is responsible for the management of the PSM program, how employees can learn about it, and how suggestions for improvement can be implemented.

On union plants, the employee representatives will be appointed by the union. On non-union plants, the employees may choose someone to represent their interests. However, this is done it is essential that the appointment be made by the employees, not management.

(2) Consultation

As already discussed, employees must be involved in all aspects of PSM, not merely informed about decisions that have been made by other people. Their opinions matter, and should always be acted on. Even when an idea is rejected, management should always communicate with the employee as to why that decision was made.

(3) Access To Information

In addition to consulting with employees, it is important that management makes sure that employees know that they have a right to access to information to do with process safety. The fact that PHA's are highlighted within the element has prompted many companies to make sure that operators participate in the PHA's, often on a rotating basis.

IMPLEMENTATION

The Employee Participation can be organized into three steps:

1. Develop and communicate a written plan.
2. Set up safety committees.
3. Establish links to other elements of PSM.

DEVELOP AND COMMUNICATE A WRITTEN PLAN

Management and the employees should develop a written plan showing how they plan to implement Employee Participation. An example of one of these is shown in Table 3-1.

Table 3-1

Representative Statement Of PSM Policy

1. The PSM program will involve all employees and contract workers in the Process Safety Management Program, as appropriate to their job function and experience level.

2. The PSM program will involve the full participation of "employee representatives" — where such duly elected representatives exist.

> 3. "Employees" includes not only full-time workers, but also temporary, part-time and contract workers, as required by law under the terms of the April 5, 1993 settlement between OSHA and the United Steelworkers Of America.
>
> 4. The Company's decision as to which kinds or classes of employees should be consulted regarding specific OSHA PSM matters will take into account factors such as job functions, experience, and their degree of involvement with PSM and the company's general background.

In order to provide documentation that the Employee Participation element is being implemented, employees can be asked to sign a form such as that shown in Table 3-2 whenever they participate in PSM activities.

Table 3-2

Representative Employee Participation Form

Date _____

I, _____, have reviewed and edited as necessary the attached document labeled

_____

Print Name _____

Signed _____

Date Recorded In Training Department _____

There are many ways that management can convey information to employees. These include:

- Safety letters
- Safety meetings
- Banners, signs and posters
- Hand-outs
- Bulletin boards
- Company newspapers
- Intranet e-mail

It is not quite so easy for employees to communicate with management. One of the most effective ways of doing this can be the plant safety committee — as long as it includes a full cross-section of employees, including those who are willing to speak their minds. There are many references to employee representatives in the OSHA standard. These would usually be on the safety committee. If the facility is non-union, it is essential that the employees' representative is selected by the employees, not appointed by management.

Another form of communication is through recognition programs. If employees do something that is out of the ordinary in order to improve safety, their contribution can be recognized by awards, which should always include a considerable amount of money.

SET UP SAFETY COMMITTEES

Safety committees provide a formal channel through which management and the employees can communicate with regard to PSM issues. Usually, the primary purpose of safety committees is to provide feedback to management on safety issues that concern the employees. As such, they serve a critical role in process safety management programs. However, they are also a means whereby workers can contribute toward achieving overall process safety goals. The committees will also be instrumental in developing safety policies and procedures. The organization of a process safety program is discussed in detail in Chapter 17.

ESTABLISH LINKS TO OTHER ELEMENTS OF PSM

No element of PSM stands alone — they all link to one another. This is particularly true with regards to Employee Participation. By itself increased participation will not improve safety; it must be integrated into all the other elements. Some of these steps are quite clear, such as having employees participate in PHA's. Others are more difficult. For example, there is often a tendency to regard Mechanical Integrity as a separate stand-alone activity that is carried out by specialists. Yet it is often the operators and maintenance workers who are the first to notice corrosion and vibration problems.

Because different people need different information, it can be useful to be prepare a table showing who should receive which documents. A example of how this could be done (for just a few items of information) is shown in Table 3-3. The shaded squares show which areas require which type of information.

Table 3-3

Assignment Of Information

	Mana-gement	Engin eering	Opera tions	Maint-enance	Contr-actor	Fire Depart-ment
MSDS	▓	▓	▓	▓		▓
Emerg-ency Training	▓			▓		▓
PHA's		▓				
Incident Investig ation	▓	▓	▓	▓		

PROBLEMS WITH EMPLOYEE PARTICIPATION

INEFFICIENCIES

Although Employee Participation is fundamental to the success of a PSM program, its application can lead to some short term problems, mostly caused by inefficiencies brought about by spreading work among a large number of people, rather than assigning it to a small number of full-time specialists. For example, rotating operators through a PHA means that the analysis will be slowed down until the new people will learn about process safety and may then use it in their on-going work. However, the newcomers may bring insights and knowledge that would not otherwise have been available to the team.

Another example of this type of problem (and opportunity) occurs when the operators are each asked to check the P&ID's for a small section of the plant. It would be much quicker to have one designer go out and do the whole job, but doing so would lose the important benefits that would be gained when the operators check their own unit line by line and valve by valve. Furthermore, the operators may be able to identify problems with the P&ID's because they know how "things really are." Ultimately, the short term inefficiencies consequent on using all the operators to perform such tasks will be more than compensated for by the gains in knowledge and understanding by all parties.

DIFFICULTIES WITH AUDITING EMPLOYEE PARTICIPATION

The topic of Employee Participation is one that is very difficult to audit because it is concerned with the spirit rather than the letter of the regulation. Although it is quite simple to have employees fill out forms (such as the one in Table 3-2), it is much more difficult to determine if the spirit of the regulation is being met. One way of doing this is to see if the employees and contract workers are involved in the implementation of all the other elements. Some questions that might help guide an audit in this area are listed below.

1. Is there an overall policy or mission statement that is perceived as being real, and that is not just some form of management fad?

2. Does the process safety program have clearly defined target and objectives — including dates when these will be met — and are the employees familiar with these targets and dates?

3. Is upper management perceived as being committed to these objectives?

4. How well is the overall strategy translated into detailed plans?

5. Do the employees understand the program and the detailed plans?

6. Are they committed to it?

CONCLUSIONS

Employee Participation lies at the heart of process safety. It is relatively easy to implement a simple paperwork program which will meet the letter of the regulations. It is much harder to achieve full participation, in which all employees and contract workers are fully involved in all aspects of process safety. But, if this can be done, the quality of all aspects of the plant operation will improve dramatically.

Chapter 4

PROCESS SAFETY INFORMATION

INTRODUCTION

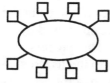 The purpose behind the Process Safety Information (PSI) element of the regulation is to make sure that the operators and other personnel have access to the information that they need in order to perform their job safely and efficiently. PSI provides the foundation for all the other elements of the process safety program. For example, Process Hazards Analyses require up-to-date and accurate P&ID's, Operating Procedures require information about process limits, and Mechanical Integrity requires information about equipment and piping.

In many cases, the information may be needed simply for reference purposes, and therefore it can be located away from the actual operation site or control room. But it is essential that the operators know that the information exists, where it is located and how it can be accessed in a timely manner.

Some of the basic questions that have to be asked before implementing the PSI program have been listed by Hoff.

1. Which documents are included?
2. Where are the documents located?
3. Who has access to them?
4. Who has the authority to change them?
5. How are revisions and changes noted?
6. Who has authority to make copies?
7. What is the retention/purge schedule?

In this context, the word "document" covers not only written information, such as is to be found in books or manuals, but also electronic images, video clips and voice recordings.

THE OSHA REGULATION

. . . The employer shall complete a compilation of written process safety information before conducting any process hazard analysis required by the standard. The compilation of written process safety information is to enable the employer and the employees involved in operating the process to identify and understand the hazards posed by those processes involving highly hazardous chemicals. This process safety information shall include information pertaining to the hazards of the highly hazardous chemicals used or produced by the process, information pertaining to the technology of the process, and information pertaining to the equipment in the process.

(1) Information pertaining to the hazards of the highly hazardous chemicals in the process. This information shall consist of at least the following:

(i) Toxicity information;

(ii) Permissible exposure limits;

(iii) Physical data;

(iv) Reactivity data:

(v) Corrosivity data;

(vi) Thermal and chemical stability data; and

(vii) Hazardous effects of inadvertent mixing of different materials that could foreseeably occur.

Note: Material Safety Data Sheets meeting the requirements of 29 CFR 1910.1200(g) may be used to comply with this requirement to the extent they contain the information required by this subparagraph.

(2) Information pertaining to the technology of the process.

(i) Information concerning the technology of the process shall include at least the following:

(A) A block flow diagram or simplified process flow diagram;

(B) Process chemistry;

(C) Maximum intended inventory;

(D) Safe upper and lower limits for such items as temperatures, pressures, flows or compositions; and,

(E) An evaluation of the consequences of deviations, including those affecting the safety and health of employees.

(ii) Where the original technical information no longer exists, such information may be developed in conjunction with the process hazard analysis in sufficient detail to support the analysis.

(3) Information pertaining to the equipment in the process.

(i) Information pertaining to the equipment in the process shall include:

(A) Materials of construction;

(B) Piping and instrument diagrams (P&IDs);

(C) Electrical classification;

(D) Relief system design and design basis;

(E) Ventilation system design;

(F) Design codes and standards employed;

(G) Material and energy balances for processes built after May 24,1992; and,

(H) Safety systems (e.g. interlocks, detection or suppression systems).

(ii) The employer shall document that equipment complies with recognized and generally accepted good engineering practices.

(iii) For existing equipment designed and constructed in accordance with codes, standards, or practices that are no longer in general use, the employer shall determine and document that the equipment is designed, maintained, inspected, tested, and operating in a safe manner.

OSHA GUIDANCE

Complete and accurate written information concerning process chemicals, process technology, and process equipment is essential to an effective process safety management program and to a process hazards analysis. The compiled information will be a necessary resource to a variety of users including the team that will perform the process hazards analysis as required under paragraph (e); those developing the training programs and the operating procedures; contractors whose employees will be working with the process; those conducting the Pre-startup reviews; local emergency preparedness planners; and insurance and enforcement officials.

The information to be compiled about the chemicals, including process intermediates, needs to be comprehensive enough for an accurate assessment of the fire and explosion characteristics, reactivity hazards, the safety and health hazards to workers, and the corrosion and erosion effects on the process equipment and monitoring tools. Current material safety data sheet (MSDS) information can be used to help meet this requirement which must be supplemented with process chemistry information including runaway reaction and over pressure hazards if applicable.

Process technology information will be a part of the process safety information package and it is expected that it will include diagrams of the type shown in Appendix B of this section as well as employer established criteria for maximum inventory levels for process chemicals; limits beyond which would be considered upset conditions; and a qualitative estimate of the consequences or results of deviation that could occur if operating beyond the established process limits. Employers are encouraged to use diagrams which will help users understand the process.

A block flow diagram is used to show the major process equipment and interconnecting process flow lines and show flow rates, stream composition, temperatures, and pressures when necessary for clarity. The block flow diagram is a simplified diagram.

Process flow diagrams are more complex and will show all main flow streams including valves to enhance the understanding of the process, as well as pressures and temperatures on all feed and product lines within all major vessels, in and out of headers and heat exchangers, and points of pressure and temperature control. Also, materials of construction information, pump capacities and pressure heads, compressor horsepower and vessel design pressures and temperatures are shown when necessary for clarity. In addition, major components of control loops are usually shown along with key utilities on process flow diagrams.

Piping and instrument diagrams (P&IDs) may be the more appropriate type of diagrams to show some of the above details and to display the information for the piping designer and engineering staff. The P&IDs are to be used to describe the relationships between equipment and instrumentation as well as other relevant information that will enhance clarity. Computer software programs which do P&IDs or other diagrams useful to the information package, may be used to help meet this requirement.

The information pertaining to process equipment design must be documented. In other words, what were the codes and standards relied on to establish good engineering practice. These codes and standards are published by such organizations as the American Society of Mechanical Engineers, American Petroleum Institute, American

National Standards Institute, National Fire Protection Association, American Society for Testing and Materials, National Board of Boiler and Pressure Vessel Inspectors, National Association of Corrosion Engineers, American Society of Exchange Manufacturers Association, and model building code groups.

In addition, various engineering societies issue technical reports which impact process design. For example, the American Institute of Chemical Engineers has published technical reports on topics such as two phase flow for venting devices. This type of technically recognized report would constitute good engineering practice.

For existing equipment designed and constructed many years ago in accordance with the codes and standards available at that time and no longer in general use today, the employer must document which codes and standards were used and that the design and construction along with the testing, inspection and operation are still suitable for the intended use. Where the process technology requires a design which departs from the applicable codes and standards, the employer must document that the design and construction is suitable for the intended purpose.

ANALYSIS OF THE REGULATION

OSHA stresses the fact that the process safety information should be written down. Not only does this mean that safety information will be available when needed, it will also help identify any hazards that may be present. The specific elements of the standard are discussed in the following sections of this chapter.

INFORMATION CONCERNING HIGHLY HAZARDOUS CHEMICALS

The term "highly hazardous chemical" is defined in 29 CFR 1910.109 and 1910.119. The term covers toxic, flammable, reactive and explosive materials. Its meaning was discussed in Chapter 2 of this book. Another definition comes from the American Society of Mechanical Engineers in ASME B31.3 — 1990 Chemical Plant &

Petroleum Piping. They define a "Category M Fluid Service" as that service in which the potential for personnel exposure is judged to be significant and in which a single exposure to a very small quantity of a toxic fluid can produce serious, irreversible harm to persons even when prompt restorative measures are taken.

MATERIAL SAFETY DATA SHEETS

Suppliers and generators of hazardous materials used in the workplace are required to document the specific hazards and related safety precautions and procedures. To do this they generally use Material Safety Data Sheets (MSDS.) MSDS have to be provided not only for the principal products and materials used in a process, but also for any chemical that may be present on site, even if the quantities are small. Therefore, a large facility may possess a library of thousands of MSDS.

The design, content and application of MSDS are explained in detail in OSHA's 1910.1200 (g) Hazard Communication Standard. Some of the key points with regard to this four page standard are as follows.

1. The manufacturer and/or distributor of a chemical is responsible for writing and distributing the MSDS.
2. The MSDS must contain information describing the hazardous properties of the chemical.
3. The MSDS must also describe flammability and explosivity properties of the chemical.
4. First aid and other treatment measures should be described.
5. A name and address of where more information can be obtained, if needed.

TOXICITY INFORMATION

Toxic chemicals can enter the human body in one of four ways:

1. Inhalation
2. Dermal (Skin) Absorption
3. Ingestion
4. Injection

It is only the first two that are usually important in process safety work. Toxic effects are affected by the following factors:

- Concentration of the material
- Duration of the exposure
- Health and sensitivity of the person(s) affected

Information on the health of each worker should also be available, particularly if he or she has been adversely affected by the working conditions. This information may include the following:

- The results of employee medical surveillance and health and epidemiology studies
- Employee and other worker health complaints
- Industrial hygiene monitoring information

EXPOSURE LIMITS

There are various methods for determining exposure limits through inhalation. Such values are often very approximate because there is not usually much information available as to how a hazardous chemical affects the human body, especially at high concentration. Generally, the only controlled tests that can be carried out are on laboratory animals. This then creates the problem as to how relevant the test results are for human beings. Moreover, the impact of even low concentrations may be difficult to evaluate because some of the health effects may take many years to emerge.

Five commonly used exposure methods are:

1. PEL-TWA — Permissible Exposure Limit, Time Weighted Average
2. TLV — Threshold Limit Values
3. IDLH — Immediately Damaging To Life Or Health
4. STEL — Short Term Exposure Limits
5. ERPG — Emergency Response Planning Guidelines

Information on these methods can be obtained from MSDS and from the NIOSH <u>Pocket Guide To Chemical Hazards</u> and from the American Conference Of Industrial Hygienists (ACGIH.)

In all cases, there is a concentration-time relationship. The higher the concentration of the chemical, the less time can a person be exposed to it before they are affected. This is shown in Figure 4-1. It can be seen that this relationship is non-linear. As concentrations increase, so permissible time exposure decreases logarithmically. In other words, the time that a person can be exposed to say 50 ppm of a chemical is less than half of the time that they can be exposed to 100 ppm.

<u>Figure 4-1</u>

<u>Concentration-Time Effects Of Hazardous Chemicals</u>

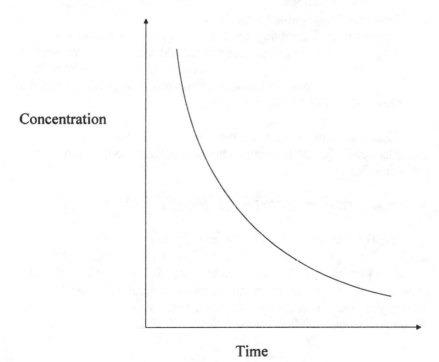

The Permissible Exposure Limit, Time Weighted Average (PEL-TWA) represents the concentration to which an average worker can be exposed to for 8 hours per day, 40 hours per week. This method is not used much in safety work because of its focus on long-term effects rather than high concentrations over a short period of time.

Threshold Limit Values, TLV, can be measured in one of three ways: TLV-TWA, TLV-STEL and TLV-C. TLV-TWA (Time Weighted Average) represents the concentration to which a worker can be exposed for 8 hours per day, 40 hours per week without suffering adverse health effects. This is equivalent to the PEL-TWA, already described.

TLV-STEL (Short Term Exposure Limit) represents the maximum concentration of the chemical to which the worker can be exposed for up to 15 continuous minutes without suffering from one or more of the following.

1. Intolerable irritation.
2. Chronic or irreversible tissue damage.
3. Narcosis of sufficient degree to increase accident proneness (narcosis affects a person's judgment, and can impair their ability to work safely.) No more than 4 excursions per day are permitted with at least 60 minutes between excursions and provided that the 8 hour TLV-TWA is not exceeded.

TLC-C (Ceiling) represents the concentration which should not be exceeded, even instantaneously. This is usually the one of most interest in safety work.

IMMEDIATELY DAMAGING TO LIFE OR HEALTH — IDLH

IDLH information is often the only form of exposure information available. It represents the effect of a 30 minute exposure to the chemical in question. IDLH values are often found in MSDS and in standard references. Some practitioners use a value of IDLH/10 as a working number for defining acceptable risk. This value is also used in some standards, such as the New Jersey TCPA.

EMERGENCY RESPONSE PLANNING GUIDELINES — ERPG

For process safety work, Emergency Response Planning Guidelines (ERPG), published by the AIHA (American Industrial Hygienists Association) are preferred. Unfortunately, not many chemicals have been assigned ERPG values yet, and so one of the other methods (usually IDLH) has to be used. To date, ERPG values are available for about 50 chemicals. There are three levels of ERPG:

ERPG 1 is the maximum airborne concentration below which it is believed nearly all individuals could be exposed for up to 1 hour without experiencing other than mild transient adverse health effects or perceiving a clearly defined objectionable odor.

ERPG 2 is the maximum airborne concentration below which it is believed nearly all individuals could be exposed for up to 1 hour without experiencing or developing irreversible or other serious health effects or symptoms that could impair their abilities to take protective action.

ERPG 3 is the maximum airborne concentration below which it is believed nearly all individuals could be exposed for up to 1 hour without experiencing or developing life threatening health effects.

Some ERPG values are provided in Table 4-1. (All values are in ppm, except where indicated)

Table 4-1

ERPG Values

Chemical Name	ERPG-3	ERPG-2	ERPG-1
1,3 Butadiene	5000	500	10
Acolein	3	0.5	0.1
Acrylic Acid	750	50	2
Allyl Chloride	300	40	3
Ammonia	1000	200	25
Benzyl Chloride	25	10	1

Chemical Name	ERPG-3	ERPG-2	ERPG-1
Bromine	5	1	0.2
Carbon Disulfide	500	50	1
Chlorine	20	3	1
Chloroacetyl Chloride	10	1	0.1
Chloropicrin	3	0.2	N/A
Chlorosulphonic Acid	30 mg/m^3	10 mg/m^3	2 mg/m^3
Chloro-tribluoroethylene	300	100	20
Crotonaldehyde	50	10	2
Diketene	50	5	1
Dimethylamine	500	100	1
Epichlorohydrin	100	20	2
Ethylene Oxide	500	50	N/A
Formaldehyde	25	10	1
Hexachlorobutadiene	30	10	3
Hydrogen Chloride	100	20	3
Hydrogen Cyanide	25	10	N/A
Hydrogen Fluoride	50	20	5
Hydrogen Sulfide	100	30	0.1
Isobutyronitrile	200	50	10
Methanol	5000	1000	200
Methyl Iodide	125	50	25
Methyl Mercaptan	100	25	0.005
Monoethylamine	500	100	10
n-Butyl Isocyanate	1	0.05	0.01
Oleum, Sulfur Trioxide, Sulfuric Acid (as sulfuric acid mist)	30 mg/m^3	10 mg/m^3	2 mg/m^3
Perfluoroisobutylene	0.3	0.1	N/A
Phenol	200	50	10
Phosgene	1	0.2	N/A
Phosphorus Pentoxide	100	25	5
Sulfur Dioxide	15	3	0.3
Tetrafluoroethylene	10000	1000	200

Chemical Name	ERPG-3	ERPG-2	ERPG-1
Titanium Tetrachloride	100 mg/m^3	20 mg/m^3	5 mg/m^3
Trimethylamine	500	100	0.1
Vinyl Acetate	500	75	5

SUBSTANCE HAZARDS INDEX (VOLATILE LIQUIDS)

With regard to hazards affecting the general public, a release of a toxic liquid is likely to be less dangerous than the equivalent release of toxic gas. Liquid spills can usually be contained, and even if they do escape outside the boundaries of the facility, measures can be taken to make sure that people do not come into contact with them. Gases, however, move quickly and are much more difficult to contain. However, some liquids vaporize at ambient conditions.

Recognizing this, some authorities (such as the states of Delaware and New Jersey and the API) require that the Substance Hazards Index (SHI) for each chemical be developed for each hazardous chemical (assuming that sufficient data are available.) This index incorporates terms for both toxicity and the ease with which the liquid can vaporize following a spill. The following definition for SHI is used.

$$SHI = \frac{EVP}{760} * \frac{10^6}{ATC}$$

EVP is the substance's vapor pressure in mm Hg at 20°C; ATC (which is the same as ERPG-3) is its acute toxicity concentration in ppm, defined as the lowest reported concentration based on recognized scientific test protocols, that can cause death or permanent injury to humans after a single exposure of one hour or less.

Although SHI is a simple way of determining the danger associated with a toxic gas, it does not cover all parameters, such as the density of the gas cloud once it is in the atmosphere. Nor is it useful in distinguishing between materials which are always present in the gas phase at all times.

PHYSICAL, REACTIVITY AND CORROSIVITY DATA

The MSDS should provide the following information on the hazardous chemicals.

- The chemical's physical status — for example, whether the phase is solid, liquid or gas.

- The chemical reactivity of the hazardous chemical, both when isolated and when mixed with other chemicals in the process, or with air that might enter the system.

- Information regarding the corrosivity of the various streams should be supplied. If an operator is provided with a new type of sample container, for example, there should be information which tells him whether the material is safe.

THERMAL AND CHEMICAL STABILITY DATA

Some chemicals may degrade exothermically if they are isolated and then exposed to heat. This may happen, for example, if a tank car of one of these chemicals is isolated on a railroad track for a number of days in the middle of summer. If the chemical reaction is suppressed through the addition of an inhibitor, it is necessary to make sure that the inhibitor does not become used up.

INADVERTENT MIXING

OSHA is particularly concerned with the effects of accidentally mixing two chemicals which may be quite harmless on their own and in the environment in which it is intended that they be contained, but which can cause a catastrophic accident if inadvertently mixed. In practice, this information will often come out of the Process Hazards Analysis.

SOURCES OF INFORMATION

In order to find about the hazardous properties of chemicals, the following can be useful sources of information:

- ANSI (American National Standards Institute)
- API (American Petroleum Institute)
- ASME (American Society of Mechanical Engineers)
- Dangerous Properties Of Industrial Materials, van Nostrand Reinhold
- Emergency Action Guides, Association of American Railroads
- NFPA (National Fire Protection Association)
- Handbook Of Chemistry & Physics, CRC Press
- Hazardous Chemical Desk Reference, van Nostrand Reinhold
- Merck Index, Merck Company
- NIOSH/OSHA Pocket Guides To Chemical Hazards
- Chemical Engineers Handbook, McGraw-Hill

INFORMATION CONCERNING TECHNOLOGY

BLOCK FLOW DIAGRAMS

Block flow diagrams provide an overall view of the process, with each major operating step represented by a block. Major process streams are shown connecting the blocks. A block flow diagram may also show a few of the more important operating parameters, such as flow rates and temperature.

Figure 4-2 is an example of a very simple block diagram. It shows a commonly found process schematic. Two feed streams enter the unit. They are purified, mixed and sent to the reaction section. The material leaving the reaction section is a mixture of product and unreacted feed materials. They go to a separation unit, where the product goes forward to purification, and the unreacted feed returns to the reaction section.

Figure 4-2

Block Flow Diagram

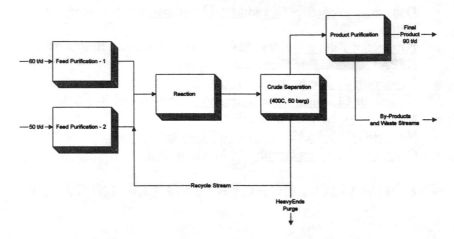

Block diagrams are used primarily for training people who are not familiar with the unit. They are also useful for conceptual safety studies because they provide a good overview of the process.

PROCESS CHEMISTRY

Knowledge of the process chemistry must be made available to those who need such information in order to carry out their work safely. It is important to make sure that the information is presented in a usable form. Most operators are not trained to understand complex chemical equations, so the process chemistry must be explained in terms appropriate to their background and training. In some cases, process chemistry may be a Trade Secret. Ways of handling this issue are discussed on page 62.

MAXIMUM INTENDED INVENTORY

Maximum intended inventory is the maximum amount of the material that can be stored onsite. Generally, this value should be as low as possible. Then, if there is an accident, the effects of a release of the chemical are minimized.

SAFE UPPER AND LOWER LIMITS

One of the most important parts of the Process Safety Information element concerns Safe Upper and Lower Limits for the major operating variables. Generally, these limits should be developed for each board-mounted instrument (or signals displayed on the DCS.) There are five values for each signal, as illustrated in Table 4-2.

Table 4-2

Operating Limits

	TRC-121 *Board-Mounted*		Units °C		
	Reactor R-101 Product Temperature Recorder and Controller				
Lo-Lo	Lo	Normal	Hi		Hi-Hi/ Interlock
—	125	175	210		225

The normal value represents the target conditions for normal operations. It will vary (within the Hi and Lo range) depending on plant conditions and customer requirements at any given time. In Table 4-2, the normal operating value for TRC-121 is 175°C.

Bracketing the normal value are the Lo and Hi values. In this example they are set at 125 and 210°C respectively. These values represent the acceptable range of safe operation. They have two important uses. First, if the operator inadvertently exceeds one of these limits, he knows that that plant is in an unsafe condition and that action must be taken to restore the value to the safe range. Second, these limits provide the allowable range for operating conditions as set by supervision and management. If a manager decides to raise the temperature of TRC-121 from 175°C to 205°C in order to increase production rates, he or she can do so knowing that this will not create a safety problem. However, if he or she wants to go outside the safe operating range, say to 220°C, then the Management of Change process must be initiated and a new value for the Safe Upper Limit must be set (after an appropriate safety analysis.)

The next set of values in Table 4-2 are Lo-Lo and Hi-Hi. If either of these values is reached, there is urgent danger and immediate action must be taken. In this example, the Hi-Hi value is 225°C. There is not a Lo-Lo value, indicating that very low temperatures do not represent a critical safety problem in this case. Often, if the Hi-Hi value is reached, the instrumentation will automatically bring the plant into a safe state.

CONSEQUENCES OF DEVIATIONS

Operators need to know the consequences should they exceed safe upper and lower limits. (Consequences are often determined from Process Hazards Analyses.) In the above example, it may be that at 210°C a high reaction rate will take place, possibly damaging the reactor catalyst and requiring a shut down of the unit. At 225°C, there may be the possibility of a run-away reaction, leading to a possible fire and explosion.

Operators should also have information, operating procedures and training as to what actions to take if a variable does go outside the permitted range.

EQUIPMENT AND ENGINEERING INFORMATION

PIPING AND INSTRUMENT DIAGRAMS

Piping & Instrument Diagrams (P&ID's) are at the heart of any process safety management program. They provide a pictorial representation of all equipment, instrumentation and piping in a facility. If the P&ID's for a unit are out of date, it is essential that they be updated immediately, because they provide the foundation for almost all the other elements of PSM. For a company that is starting a new process safety program, updating the P&ID's should be one of the first tasks to be carried out. Otherwise, all other elements of PSM will be built on insecure foundations.

Information that is typically to be found on a P&ID includes the following:

- Piping and valves (including sizes)
- Equipment (with principal dimensions)
- Instruments
- Control Loops
- Materials of construction
- Insulation

The following information is not usually put on the P&ID:

- Process information (flows, temperature, pressures)
- Distances (P&ID's are not to scale)

When checking or validating P&ID's, good practices include the following:

- They should not be too crowded. A crowded P&ID is difficult to follow. It is also very hard to make further additions.
- They should be neat. Since almost all P&ID's are now generated by CAD software, neatness is not hard to achieve. Neatness makes the drawings more usable.
- They should be simple.
- They should be consistent. Generally, this is achieved by using the first P&ID in a series — the Legend Sheet — to define the symbols and conventions used on this particular process (such as line labeling conventions.)
- They must be accurate. This is really the most important item of all. If they contain many errors, users will lose confidence in them, and they will not be used.

In order to indicate their status, P&ID's will often have a code letter in front of the number. For example, the letter 'W' means working, i.e. the P&ID is still being checked or developed. The letter 'I' may mean that the drawing has been issued for construction and that it has received all the necessary engineering checks, and can be used for turnover packages (see Chapter 9.) Finally, the letter 'A' can be used to indicate "as-built"; this version of the P&ID shows what is actually in the field.

When P&ID's are updated, there should be some method of quickly identifying which items have changed. Often this is done by "clouding" the changed sections.

MATERIAL AND ENERGY BALANCES

Material and Energy balances are normally shown on the Process Flow Diagrams (PFD's), which are a development of the block diagrams already discussed. A PFD contains process information for all significant streams. This information typically includes flow rates, phases, temperatures, pressures, viscosities, thermal conductivities and specific heats. With the PFD, a complete heat and material balance for the unit can be obtained.

PFD's are not usually used much once a plant is built. They are too complex to provide a simple overview, such as is obtained from the Block Flow Diagrams. Nor do they contain mechanical information, so they have limited value in the field. Furthermore, once a plant is built operating conditions are usually changed (often to obtain higher production rates.) It is unusual for a facility to keep the PFD's up to date to reflect this kind of change (nor does OSHA require that they do so.)

RELIEF SYSTEMS

The information provided concerning relief valves should include worst case scenarios and the calculations that were made for the design of headers and flares. Relief valves are to be sized using the NFPA methodology specified in 29 CFR 1910.106. The design of relief valves for complex, two-phase flow systems is discussed in literature from the Center for Chemical Process Safety.

DESIGN CODES AND STANDARDS

It is useful to have available the Design Codes and Standards that were used when the plant was new. For those plants that are more than a few years old, it will often be found that the standards that were used when it was being designed are no longer current, but normally this does not mean that the equipment has to be changed; it can be "grandfathered." However, if a change is made to the process that requires a redesign, then the current codes must be used, even on the old equipment.

SAFETY SYSTEMS

In many plants, alarms and interlocks are critically important to safety. They warn the operators that the plant conditions are becoming unsafe. They also provide a capability for the plant to automatically take over if the operator does not respond properly to an incipient emergency condition for any reason.

Many plants, particularly those with Distributed Control Systems (DCS), have three types of alarm. The first is an operating alarm; it is used to assist with operations and is not concerned with safety. For example, an operator may be preparing a solution of polymerization inhibitor. The recipe for this may change from batch to batch. If the tank for a particular batch is to be filled to say the 65% level, the operator can set the alarm at that value, start filling the tank and then move on to some other activity. When the alarm sounds, he or she will stop the fill process. If the next batch requires a 75% level, the operator can adjust the alarm correspondingly.

The second type of alarm tells the operator that the plant is moving out of its pre-defined safe operating range (the Lo and Hi values shown in Table 4-2.) In the case of the fill tank described in the previous paragraph, the Hi alarm value may be set at say 95%. It tells the operator that, if the filling process continues above that point, the tank will soon overflow, thus creating a hazardous condition. The corresponding Lo alarm may be set at 10% in order to protect a pump on the base of the tank from cavitating. These alarm settings cannot be changed without going through the Management Of Change process.

It is important not to have too many alarms of this type, otherwise, during an emergency, the operators may become overloaded with information, and hence they may not be able to determine what is going on because of the amount of information that they have to process. This problem of information overload was an important factor in the Three Mile Island nuclear accident.

The third type of alarm — which is often linked to an automatic shutdown system — is sometimes referred to as a "hard" alarm, and has its own dedicated circuitry. This alarm system protects the plant against major catastrophes. The instrumentation is typically of very high reliability, and the settings cannot be changed by the operations personnel. Needless to say, an extremely thorough safety review and MOC analysis must be performed before these critical alarm values can be changed.

INTERLOCKS AND TRIPS

The words *interlock* and *trip* are often used interchangeably. Strictly speaking, however, they are different. An interlock prevents an action from being taken, whereas a trip causes an action to be taken. Interlocks are commonly found on hazardous systems such as oxygen mix stations. They prevent oxygen from being added to the system unless the correct concentration of hydrocarbon is present. Trips shut down systems that are already in operation. For example, a high temperature trip on a furnace will shut off all the burners. In the example shown in Table 4-2, the Hi-Hi alarm has a trip associated with it so that, even if the operator did not take action, the system would fail in a safe condition.

Most trips are activated by a signal from the plant. However, some trips are manually activated. For example, a "panic" button on the board can be pressed to shut down the whole unit quickly and safely.

EMERGENCY ISOLATION VALVES

If there is a serious fire on a unit, it may be impossible for the operators to reach the valves that must be closed to stop additional flammable material from feeding the fire. Therefore, it is useful to have Emergency Isolation Valves (EIV's) at the unit's perimeter. If these valves are closed, the flow of all hazardous chemicals into the unit will be stopped and the fire will go out once the inventory of material within the unit has been consumed. The EIV's must be located and protected such that (a) they are not damaged in a fire or explosion, and (b) operators can reach them in an emergency. EIV's can be either manual or automatic. However, if they are automatic it must also be possible for the operators to reach them in an emergency and to close them by hand.

CONCLUSIONS

Accurate, useable and complete process safety information is a vital part of any process safety program. Developing the information often takes a lot of time and effort. In particular, the development of safe upper and lower limits can be a difficult and time-consuming task. However, without a sound information base, the quality of the remainder of the process safety program will always be in doubt. Therefore it is critical that this element be given high priority, particularly when a new process safety program is being started.

Chapter 5

PROCESS HAZARD ANALYSIS

INTRODUCTION

 The purpose of a Process Hazard Analysis (PHA) is to identify hazards, and to provide a preliminary estimate as to how serious they might be and what actions can be taken. Originally, PHA's were conducted on facilities that were still in the design stage (Lawley, 1974.) If any major hazards were uncovered there was still time to change the process itself. Since then, PHA's have been used increasingly for analyzing facilities that are already operating. In these cases, identified hazards are normally addressed by adding safety features, or by improving administrative items, such as operating procedures and training.

Risk consists of three elements: the hazard, its consequence and the probability of it actually happening (these topics are discussed more fully in Chapter 16.) Although PHA's focus on the first of these — hazard identification — they can also offer preliminary insights into consequence and frequency issues.

THE OSHA REGULATION

(1) The employer shall perform an initial process hazard analysis (hazard evaluation) on processes covered by this standard. The process hazard analysis shall be appropriate to the complexity of the process and shall identify, evaluate, and control the hazards involved in the process. Employers shall determine and document the priority order for conducting process hazard analyses based on a rationale which includes such considerations as extent of the process hazards, number of potentially affected employees, age of the process, and operating history of the process.

[Schedule details follow . . .]

(2) The employer shall use one or more of the following methodologies that are appropriate to determine and evaluate the hazards of the process being analyzed.

(i)   What-If;

(ii) Checklist;

(iii) What-If/Checklist;

(iv) Hazard and Operability Study (HAZOP);

(v)   Failure Mode and Effects Analysis (FMEA);

(vi) Fault Tree Analysis; or

(vii) An appropriate equivalent methodology.

(3) The process hazard analysis shall address:

(i) The hazards of the process;

(ii) The identification of any previous incident which had a likely potential for catastrophic consequences in the workplace;

(iii) Engineering and administrative controls applicable to the hazards and their interrelationships such as appropriate application of detection methodologies to provide early warning of releases. (Acceptable detection methods might include process monitoring and control instrumentation with alarms, and detection hardware such as hydrocarbon sensors.);

(iv) Consequences of failure of engineering and administrative controls;

(v) Facility siting;

(vi) Human factors; and

(vii) A qualitative evaluation of a range of the possible safety and health effects of failure of controls on employees in the workplace.

(4) The process hazard analysis shall be performed by a team with expertise in engineering and process operations, and the team shall include at least one employee who has experience and knowledge specific to the process being evaluated. Also, one member of the team must be knowledgeable in the specific process hazard analysis methodology being used.

(5) The employer shall establish a system to promptly address the team's findings and recommendations; assure that the recommendations are resolved in a timely manner and that the resolution is documented; document what actions are to be taken; complete actions as soon as possible; develop a written schedule of when these actions are to be completed; communicate the actions to operating, maintenance and other employees whose work assignments are in the process and who may be affected by the recommendations or actions.

(6) At least every five (5) years after the completion of the initial process hazard analysis, the process hazard analysis shall be updated and revalidated by a team meeting the requirements in paragraph (e)(4) of this section, to assure that the process hazard analysis is consistent with the current process.

(7) . . . .

OSHA GUIDANCE

A process hazard analysis (PHA), sometimes called a process hazard evaluation, is one of the most important elements of the process safety management program. A PHA is an organized and systematic effort to identify and analyze the significance of potential hazards associated with the processing or handling of highly hazardous chemicals. A PHA provides information which will assist employers and employees in making decisions for improving safety and reducing the consequences of unwanted or unplanned releases of hazardous chemicals. A PHA is directed toward analyzing potential causes and consequences of fires, explosions, releases of toxic or flammable chemicals and major spills of hazardous chemicals. The PHA focuses on equipment, instrumentation, utilities, human actions (routine and non routine), and external factors that might impact the process. These considerations assist in determining the hazards and potential failure points or failure modes in a process.

The selection of a PHA methodology or technique will be influenced by many factors including the amount of existing knowledge about the process. Is it a process that has been operated for a long period of time with little or no innovation and extensive experience has been generated with its use? Or, is it a new process or one which has been changed frequently by the inclusion of innovative features? Also, the size and complexity of the process will influence the decision as to the appropriate PHA methodology to use. All PHA methodologies are subject to certain limitations. For example, the checklist methodology works well when the process is very stable and no changes are made, but it is not as effective when the process has undergone extensive change. The checklist may miss the most recent changes and consequently the changes would not be evaluated. Another limitation to be considered concerns the assumptions made by the team or analyst. The PHA is dependent on good judgment and the assumptions made during the study need to be documented and understood by the team and reviewer and kept for a future PHA.

The team conducting the PHA need to understand the methodology that is going to be used. A PHA team can vary in size from two people to a number of people with varied operational and technical backgrounds. Some team members may only be a part of the team for a limited time. The team leader needs to be fully knowledgeable in the proper implementation of the PHA methodology that is to be used and should be impartial in the evaluation. The other full or part time team members need to provide the team with expertise in areas such as process technology, process design, operating procedures and practices, including how the work is actually performed, alarms, emergency procedures, instrumentation, maintenance procedures, both routine and non routine tasks, including how the tasks are authorized, procurement of parts and supplies, safety and health, and any other relevant subject as the need dictates. At least one team member must be familiar with the process. The ideal team will have an intimate knowledge of the standards, codes, specifications and regulations applicable to the process being studied. The selected team members need to be compatible and the team leader needs to be able to manage the team and the PHA study. The team needs to be able to work together while benefiting from the expertise of others on the team or outside the team, to resolve issues, and to forge a consensus on the findings of the study and the recommendations.

The application of a PHA to a process may involve the use of different methodologies for various parts of the process. For example, a process involving a series of unit operations of varying sizes, complexities, and ages may use different methodologies and team members for each operation. Then the conclusions can be integrated into one final study and evaluation. A more specific example is the use of a checklist PHA for a standard boiler or heat exchanger and the use of a Hazard and Operability PHA for the overall process. Also, for batch type processes like custom batch operations, a generic PHA of a representative batch may be used where there are only small changes of monomer or other ingredient ratios and the chemistry is documented for the full range and ratio of batch ingredients. Another process that might consider using a generic type of PHA is a gas plant. Often these plants are simply moved from site to site and therefore, a generic PHA may be used for these movable plants. Also, when an employer has several similar size gas plants and no sour gas is being processed at the site, then a generic PHA is feasible as long as the variations of the individual sites are accounted for in the PHA. Finally, when an employer has a large continuous process which has several control rooms for different portions of the process such as for a distillation tower and a blending operation, the employer may wish to do each segment separately and then integrate the final results.

Additionally, small businesses which are covered by this rule, will often have processes that have less storage volume, less capacity, and less complicated than processes at a large facility. Therefore, OSHA would anticipate that the less complex methodologies would be used to meet the process hazard analysis criteria in the standard. These process hazard analyses can be done in less time and with a few people being involved. A less complex process generally means that less data,

P&ID's, and process information is needed to perform a process hazard analysis.

Many small businesses have processes that are not unique, such as cold storage lockers or water treatment facilities. Where employer associations have a number of members with such facilities, a generic PHA, evolved from a checklist or what-if questions, could be developed and used by each employer effectively to reflect his/her particular process; this would simplify compliance for them.

When the employer has a number of processes which require a PHA, the employer must set up a priority system of which Phase to conduct first. A preliminary or gross hazard analysis may be useful in prioritizing the processes that the employer has determined are subject to coverage by the process safety management standard. Consideration should first be given to those processes with the potential of adversely affecting the largest number of employees. This prioritizing should consider the potential severity of a chemical release, the number of potentially affected employees, the operating history of the process such as the frequency of chemical releases, the age of the process and any other relevant factors. These factors would suggest a ranking order and would suggest either using a weighing factor system or a systematic ranking method. The use of a preliminary hazard analysis would assist an employer in determining which process should be of the highest priority and thereby the employer would obtain the greatest improvement in safety at the facility.

Detailed guidance on the content and application of process hazard analysis methodologies is available from the American Institute of Chemical Engineers' Center for Chemical Process Safety.

ANALYSIS OF THE REGULATION

(1) Initial Hazard Analysis

The order in which PHA's are carried out is usually determined by a preliminary consequence ranking, i.e. those processes which have the potential for the most serious consequences are analyzed first. In practice, the ordering of PHA's is also affected by logistical issues such as the availability of up-to-date P&ID's or of key personnel.

(2) Methodology

The OSHA standard lists six different methods for conducting a PHA. It also provides for the use of other methods that are not listed in the regulation. (The various methods are described later in this Chapter.) The key point is that the regulation permits a great deal of flexibility regarding the choice of method. In particular, there is no requirement that the HAZOP method be used.

(3) Issues To Address

In addition to identifying process hazards, the PHA must review previous incidents which could have caused a serious accident. (The analysis of incidents is discussed in Chapter 13 — Incident Investigation.) There is also a requirement that the causes of near-misses be addressed. The standard identifies administrative issues (such as operating procedures and training) as being an important of a PHA. It also specifically calls out siting and human factors. These two topics are discussed later in this Chapter.

(4) Team

One of the key elements of a PHA, regardless of which method is chosen, is that it be performed by a team (with the possible exception of fault tree analysis and those methods that use rigorous and extended logic — it may be better if these are carried out by just one analyst.) Team members should represent the various operations, technical disciplines and contractor groups present in the facility or on the design team. The creation of a representative team helps address the Employee Participation requirements of the standard.

(5) Findings And Recommendations

It is likely that the PHA will identify hazards that require attention. The standard requires that these situations be addressed in an expeditious manner, otherwise the facility may be cited for a willful violation.

(6) Revalidation

OSHA requires that the PHA be updated and revalidated at least every five years. The regulation does not require that a full PHA be carried out to do this — although it may make sense to do so if there have been a lot of changes to the facility. Therefore, what is needed is a method that might be termed "PHA-by-exception." This is a way of examining the full PHA that was performed originally, and checking it out against the changes that have been made since it was issued. If the other elements of PSM have been properly implemented, particularly Management of Change and Prestartup Safety Reviews, this PHA validation should be quite straightforward. If the first PHA was a HAZOP, it is often appropriate for the subsequent analyses to use either the What-If or the Checklist methods. This will save time, and will probably provide for a superior analysis because people will be able to retain their freshness.

SEQUENCE OF PHA'S

When a new process is being designed and constructed, it is normal for PHA's to be performed at each stage of the design. The choice of method will differ at each stage, reflecting the increasing amount of engineering data that is available. These stages in PHA development are discussed below.

CONCEPTUAL PHA

A conceptual PHA is used to provide a preliminary safety analysis while the basic process or plant design is still being developed. For example, one company had developed a process that had good economics but that used large quantities of a very hazardous chemical. The chemical had very little odor, making it difficult to detect in the event that it were to be released. Only when a reliable instrument "sniffer" that could detect the gas was developed did senior management give the go-ahead to commercialize the process.

A conceptual PHA provides an excellent opportunity for eliminating hazards entirely, and for making fundamental changes to the process to achieve an Inherently Safer Design. For example, a hazardous catalyst may be replaced with one that is much less hazardous. Or it may be found that some equipment can be entirely eliminated, thus reducing that particular risk to zero ("if it's not there it can't leak.") Also a conceptual PHA can explore the use of lower temperatures and pressures in all sections of the process. This is definitely the best time to make changes to the process because all changes are on paper at this stage; no equipment has been purchased. Conceptual PHA's may also identify critical missing information, such as flammability information on process chemicals.

At this stage in the design there is no detailed engineering documentation. Therefore those methods such as HAZOP that require completed P&ID's are not suitable. Nor is a checklist approach likely to be of much benefit because there is no plant experience on which to draw (unless the plant is based on a previous design.) Therefore, the What-If method is probably the best method for this phase of the PHA because it is good at encouraging conceptual and creative thinking.

PRELIMINARY DESIGN

Once the basic process design is complete, a Preliminary Design PHA can be carried out. The available documentation will generally be limited to block flow diagrams, some preliminary Process Flow Diagrams (PFD's) and Material Flow Diagrams (which provide information on materials of construction.) Once more, the What-If method is a good choice because there are still opportunities to make major changes in the process design and to reduce inventories of hazardous chemicals.

The What-If/Checklist method is also good at this stage. The What-If approach encourages broad-range thinking while the Checklist questions provide a framework on which to base the analysis.

FINAL DESIGN

At the conclusion of the final design, a complete set of P&ID's will be available. Other documentation that should be available to the PHA team includes electrical loop drawings, MSDS and draft operating manuals. This is the stage that is most appropriate for a full Hazard and Operability study (HAZOP.)

EXISTING PLANTS

As has already been discussed, PHA's were first developed to assist designers to analyze new plants for safety and operability problems, most PHA's are actually performed for units that are already constructed and operating. A PHA for an existing unit differs from a PHA for a new facility in three important ways. First, there is actual operating experience to draw on, thus making the Checklist method a good choice. Second, because the plant is already built, any changes are likely to be much more expensive and more difficult to execute than they would be for a new design. The third difference is to do with the team. Their experience will help identify known problems, but it may make it harder to think of what might be. Statements such as "I've been here 12 years, and I've never seen that happen" contain the implicit rider: "and therefore it cannot happen." This is not a problem with PHA's conducted on brand new plants.

PHA's are often included in the Management of Change (MOC) process. For example, if it is planned to change a piece of equipment, the MOC and Prestartup Safety Review processes can call for a PHA to be carried out to determine if the change could cause any safety problems. Such PHA's are often quite brief because the proposed change could be quite minor. Often, a What-If or Checklist approach (rather than a full HAZOP) is sufficient.

One issue that the team should discuss at the start of a PHA of a unit that is already operating is how to analyze unused or abandoned equipment that is still present on the site. An unused storage tank, for example, may have been declared as abandoned. Yet, if that tank can be filled with process liquids merely by opening valves, then it is not really abandoned, and so it has to be included in the analysis. True abandonment of equipment only occurs when the equipment is positively isolated and/or disconnected (including all associated instrumentation and electrical equipment.)

DECOMMISSIONING / DEMOLITION

When a plant is decommissioned, it has two possible fates. The first is that it will be simply mothballed, so that it can be restarted at some unspecified time in the future when economic conditions improve. The second possibility is that the plant will be torn down and the site used for something else. In either case, a PHA should be performed, with the What-If technique probably being the preferred method. In the case of the plant that is being mothballed, the analysis will include items such as the following:

- Ensure that rotating equipment is turned on a regular basis.
- Check for leaks into and around equipment.
- Check electrical and instrument systems for integrity.

If the plant is to be demolished, the checklist will focus on items such as:

- Hidden pockets of hazardous chemical in the equipment and piping.
- Contaminated soil.
- Hazardous materials construction materials such as asbestos insulation.

ORGANIZATION OF A PHA

STEERING COMMITTEE

Many PHA teams are sponsored by a management group, which often has a title such as "Steering Committee." The PHA team leader reports to this committee. The committee will include representatives from operations, maintenance and the risk management group. The committee has responsibilities such as the following:

1. Provide the team with a Charge Letter, or equivalent, as discussed below.

2. Establish that the links with other PHA's are properly made. PHA's on large facilities usually cover only a relatively small fraction of the total process. It is important to make sure that the connections between the different processes are properly analyzed, particularly with regard to plant utilities.

3. Establish policy changes. For example, one PHA team was analyzing a unit that included an oxygen mix station, in which hydrocarbon vapor and pure oxygen are mixed together. This was a well-known and well-understood high risk scenario which had already been analyzed in depth many times using a variety of techniques. Consequently, the Steering Committee agreed that the PHA team could exclude the oxygen mix station from its scope of work.

CHARGE/SCOPE LETTER

Management should issue a charge letter before the PHA starts. The letter should list the basic parameters of the PHA, including items such as:

- *Objective.* The Charge Letter should define what it is that the team is expected to achieve. For example, a team may be charged with identifying only major safety problems. Minor safety or health issues are declared to be outside the scope of the project, as are operability and reliability issues (although drawing a clear line between minor and major problems is difficult.)

- *Physical Scope.* The area to be analyzed is usually a discrete process unit, such as "The Secondary Reaction Section" or "Boilers #1 and #2." Such units usually have multiple connections with other areas, and often with other companies (via pipelines and truck deliveries, for example.) It is very important to define these boundaries and to decide how the hazards associated with these interfaces are to be analyzed.

- *Method(s) To Be Used.* The charge letter will describe which PHA methods are to be used for this project. It may, for example, specify that HAZOP is to be used for all process units, but that Checklist is satisfactory for all batch operations, and that Failure Modes and Effects Analysis (FMEA) can be used for the analysis of machinery such as turbines and motors.

- *Personnel.* The charge letter will specify who is to be on the PHA team. Generally, specific individuals will be named. However, it may be enough simply to specify a job function, such as "Operations Specialist."

- *Risk Management Guidance.* If the team is expected to develop preliminary risk management analyses, this should be spelled out in the charge letter, along with the methods to be used.

- *Schedule.* The charge letter should provide an estimate as to how much time it is expected that the PHA will take (including any training that may be necessary.)

PREPARATIONS

One of the keys to a successful PHA is that all the team members, particularly the leader and scribe, are thoroughly prepared. Indeed, the importance of this step cannot be overstated. When PHA's go badly, the reason is usually that the team leader and scribe were not fully prepared. Hence, the PHA got off to a bad start from which it never recovered. (Another way in which PHA's can go badly is to do with the writing of the final report. This has to be done quickly.) As a working number, it is good to budget half a day of preparation time for each meeting day, and another half day per meeting day for writing the report.

Every member of the team should have reviewed the P&ID's and other documentation carefully before the analysis starts, and each person should already have some suggestions for the upcoming discussions. The scribe should make sure that logistical issues such as the following have been taken care of.

1. Has the room been reserved?
2. Is it big enough?
3. Are drinks, snacks, lunch to be provided?
4. If so, have they been ordered, and is there a charge number for them?
5. Is a computer available?
6. Is a projector available (if needed)?
7. Have the participants been provided with a "package", containing the charge letter, shot-down P&ID's, copies of key MSDS and other pertinent materials?

8. Is there an ample supply of pens and paper (including highlighters to mark the drawings)?
9. Is there space on the wall to attach full-size P&ID's?

KICK-OFF MEETING

At the start of the PHA, it is a good idea for a senior manager to lead a kick-off meeting. He or she will explain the company's commitment to the PHA process, and describe some of the details of the charge letter.

DOCUMENTATION

Because PHA's take up the time of many valuable people, it is important to carry out as much work and to collect as much information as possible before the meetings start so that the team can concentrate on the analysis itself. The most critical task is to make sure that a complete and current set of P&ID's is available. If it is not, the PHA should probably be postponed because it can be very frustrating trying to work around missing or unreliable information. One set of P&ID's should be designated as "official." This contains all the comments and marks representing the team discussions. This set is also the one that is kept on file in the event that there is an incident investigation or an audit following the completion of the HAZOP. Once the PHA meetings start, the official set of P&ID's is often maintained by the process engineer on the team. Neither the scribe nor the team leader should have this responsibility. The scribe is already very busy just making sure that the notes are accurate and representative of the discussion. The team leader needs the freedom to concentrate on stimulating the discussion and encouraging thinking. This does not leave sufficient time to mark up the P&ID's.

Other preliminary materials usually needed for the PHA include:

- Operating procedures
- Safe upper and lower limits (particularly pressure and temperature limits for the equipment items)
- MSDS
- MAWP (maximum allowable working pressure) for vessels
- Piping specs

- Process Chemistry
- Lower flammable and explosive limits
- Electrical area classification plot plan

LOCATION OF THE PHA

Many larger facilities find that PHA's are going on almost all the time. In such situations, it is good to have a dedicated PHA room which will contain built-in facilities such as an overhead projector, a library of risk management materials and a large coffee pot (with coffee in it.)

MEETING RULES

It is important to make sure that all normal meeting rules are observed during a PHA. These rules include:

- *Respect.* The team members should respect one another's points of view, particularly when there is disagreement.

- *Punctuality.* If a person is late for a meeting, or leaves early, then they have wasted not only their own time, but the time of everyone else at the meeting. This is discourteous and inefficient.

- *Priority.* PHA's often take a long time and they require the involvement of key personnel. There is often considerable pressure for these people to skip a session or two so that they can get back to do their "real work." It should be made clear that a PHA is real work. This means that they should not take telephone calls or answers beepers and radios during the meetings.

- *Time.* PHA's can take a considerable amount of time. A general rule of thumb is that the meetings should not go on for more than six hours a day, otherwise the participants will lose their concentration. Not all PHA's have to take a long time. If a team is analyzing a small Management of Change problem, for example, the PHA itself may take just a few minutes. For these PHA's, the biggest problem can be making sure that the team's findings are properly documented.

- *Focus On The Leader.* All conversation and comments should be directed through the team leader. Side conversations can be very distracting to the other team members.

THE FINAL REPORT

A PHA is really just a particular form of audit — one in which the auditors create ways in which accidents might happen. Therefore, the end product of a PHA — like any other type of audit — is a report. This must be done well, and it must be done quickly. It is not an "extra"; the report is a vital document. Yet many PHA reports are not well done. In particular, they tend to be much too cryptic. When the report is read say six months after the PHA, the reader wants to know what was going on during the discussions. In particular, the reader wants to know what was said about each of the recommendations. In fact, it is sometimes startling to see how cryptic and opaque PHA reports are. Even the original team members cannot understand what is written down.

An example Table Of Contents for a PHA report is provided in Table 5-1.

Table 5-1

Representative Table Of Contents For A PHA Report

A.   Executive Summary
    A.1   The Purpose Of The PHA
    A.2   Regulatory Issues
    A.3   How It Was Done
    A.4   Who Was Present
    A.5   "Showstopper" Recommendations
B.   Recommendations
C.   PHA Notes

For most executives, there are two critical items to do with the report. The first is a statement concerning what regulations are in place, and how this PHA addressed those recommendations. The second critical item concerns those recommendations that are either very expensive, or that could radically change the nature of the project.

Part B of the report ("Recommendations") should list all of the recommendations that were generated, along with as much detail as possible explaining everything to do with those recommendations. This is one reason it is important to write the report quickly. No matter how good the PHA Notes may be, a lot of information is in peoples' heads, and this information should be captured as soon as possible. For the same reason, the team members must review the report immediately. In practice, the report/review process should normally be complete within 48 hours of the PHA's conclusion. For a long PHA (more than say three days), the leader should be writing the report on an on-going basis.

STEPS OF A PHA

PHA's have five principal steps:

1. Define the purpose of the node.
2. Identify the hazards.
3. Identify the safeguards.
4. Rank of the hazards, to determine the order in which they should be addressed.
5. Develop recommendations to correct the identified hazards.

DEFINE THE PURPOSE OF THE NODE

Before the analysis starts, the nodes should be selected and the purpose each explained to the team. Typically, the process engineer will provide a brief overview as to what is taking place at the node. Operations and maintenance experts will then provide some history and operating experience to do with it. Any relevant documentation to do with that node, such as equipment data sheets or MSDS, should be put before the team at this time.

IDENTIFICATION OF HAZARDS

The identification of hazards is the PHA team's primary — and sometimes only — responsibility. The team is, in effect, carrying out an audit/design review of the unit based on hypothetical incidents. They are looking for ways in which things might go wrong; they are not concerned with the implications of their findings or how to solve the problems that have been identified.

IDENTIFICATION OF SAFEGUARDS

While identifying hazards, many teams will also consider the safeguards. For example, the hazard of high pressure may have the associated safeguards of pressure alarms and relief valves. The hazard of a leak at a pump seal has the safeguard of a remote pump shut down.

A safeguard is generally considered to be an item that is installed in addition to normal control systems and administrative procedures, and whose function is just to enhance safety. For example, a relief valve would be considered a safeguard, whereas a normal operating pressure controller would not. Similarly, special training in how to handle a particular high-hazard scenario could be regarded as a safeguard. However, the normal plant training program is not a safeguard. Post-accident safety systems, such as the existence of an Emergency Response Team or the use of foam systems to suppress fire, are not safeguards because they come into use after the event has occurred.

Safeguards can themselves create a hazard, although one that is much less serious than the one that they are protecting against. For example, a relief valve that discharges to atmosphere protects against vessel rupture, but it may also may lead to an employee being affected by the fumes.

RANKING OF HAZARDS

Once hazards have been identified, some means of ranking them is needed. For each hazard, there should be corresponding values for consequence and frequency. The performance of a formal risk analysis is outside the scope of a normal PHA team. However, the team may be able to develop some preliminary values for consequence and frequency. This will allow them to develop risk matrices (*see* page 378.) It will also provide a starting point should the company decide to carry out a formal analysis.

Formal risk ranking is a time-consuming activity that frequently requires the services of a variety of different specialists. However, it is possible for the PHA team to make a first estimate as to the risk of the hazards that it has identified. This type of preliminary screening is subjective, but it does provide some guidance as to which hazards to address first. For example, each hazard may be given a simple A/B/C risk ranking, where 'A' represents a very serious situation that has to be addressed right away, 'B' covers most of the normal findings, and 'C' represents the "nice to do" issues, along with operability and reliability problems. This is done by having the team come up with rough assessments of consequence and likelihood, and then complete a simple risk matrix. At this stage in the analysis, a 3 x 3 matrix is sufficiently complex.

At this stage, the team may also be able to identify the "low-hanging fruit" recommendations, i.e. those ideas that can lead to a quick improvement in safety without requiring a significant investment in time or money. For example, if it is found that a certain operating procedure is not properly formatted, it is easier just to fix it, rather than go through any type of risk ranking. Similarly, if a safety sign is unreadable, it should simply be replaced. (The only caution to be made about this type of recommendation is that such problems may be symptomatic of systemic management problems. For example, the PHA team may wish to ask why the system allowed the safety sign to become unreadable in the first place.)

A great advantage of having even a brief but formal risk ranking is that it introduces objectivity into an activity that can be subjective and even emotional. Too often, PHA teams work on hunches and gut feel. In particular, on plants that have been operating for some years, there can be a problem with a person who has extensive experience of the process. On the one hand, he or she may downplay some recommendations on the grounds that "I've been here for fifteen years, and I've never seen that" (with the implication that it never can happen.) Alternatively, he may say: "I had this happen to me once (from which the inference is drawn that the scenario is of high probability.)

A formal risk ranking also helps address the problem of fixation. In complex situations, people tend to fixate on one or two factors, and then exclude all other information, regardless of its relevance (fixation was an important part of the Three Mile Island nuclear power plant incident.)

Formal risk ranking may also reduce the number of recommendations. PHA teams have a tendency to be very conservative and to generate a recommendation for every identified hazard, regardless of the actual risk associated with that hazard. In effect, the PHA generates wish lists based on hunch and speculation. Formalizing the risk cuts out those recommendations that are really not justifiable.

DEVELOPMENT OF RECOMMENDATIONS

As has already been discussed, risk consists of three components: the hazard, its consequence and its likelihood. Therefore, recommendations generally take on one of three forms:

1. Remove the hazard
2. Reduce the consequence of the hazard
3. Reduce the probability of the hazard occurring

Of these, the first one — remove the hazard — is usually the best because it is the only one that can reduce risk to zero. However, it is often impossible to do.

Some recommendations are best handled by outside experts. For example, if the team has found that a certain compressor represents a hazard, the recommendation as to what needs to be done may need to come from the compressor manufacturer. For this reason, care should be taken about giving the PHA team the responsibility for fully resolving recommendations. If it does have that responsibility, then the leader should understand the importance of reaching outside the team for specialist help. PHA teams are made up of people who are expert on the unit being analyzed, but who are not specialists in specific areas of technology. The team members may simply not have the knowledge to be able to find the right solution, and they will be less efficient than the experts.

Another reason for not having the PHA team develop recommendations is that the mental process for finding hazards is quite different from that for solving problems. When finding hazards, the team has a fault-finding attitude, but when generating recommendations, the team is in a constructive, problem-solving mode. The two thought processes can be difficult to integrate in a single meeting. (It is analogous to the difference between consulting and doing, as discussed in Chapter 17.)

FORM OF THE RECOMMENDATIONS

Careful attention should be paid to the way in which recommendations are written up. If a PHA report makes a strong, unequivocal recommendation, management is faced with the possibility of undesirable attention, particularly should an incident that relates to the recommendation actually occur. For these reasons, most PHA reports use words such as "consider" and "investigate" when listing the recommendations. (Similarly, if a problem could not be found, it is better to say "no problem identified" rather than "no problem" because the second phrase suggests that there are no hazards, whereas the first recognizes that hazards do exist, but this particular team was unable to find them.)

However, not all managers agree with the need to be so circumspect. Once a hazard has been identified as being serious enough to require action, then words that soften the recommendations will not make any difference if there is an incident; the company will still be faced with a situation where a hazard was known about but had not been addressed. In this view, fine tuning the language of a report will not really make any difference.

Sometimes, management may choose to reject a recommendation if it is judged to be erroneous or infeasible (Chemical Manufacturers Association, 1993.) If management does reject a PHA recommendation, they need to show that the team did not have sufficient information at the time of the analysis, or that they misjudged a particular situation. It is not sufficient to arbitrarily cancel or postpone a recommendation — the decision must be backed by rational analysis. Guidance is also provided in OSHA's CPL 2-2.45.

ACTION ITEMS

Action Items represent problems that the team found, but which did not result in a recommendation, often because additional research or investigation is needed. If the research indicates that there is indeed a problem, then a formal recommendation can then be made. Action Items may be written in the form of a nascent recommendation, i.e. they are a recommendation with a qualifying statement saying that more information is needed before the recommendation is firm.

FOLLOW UP AND CRITIQUE THE PHA

Once the PHA is finished, it is useful to have a critique of the PHA itself in order to determine if it was well-managed and whether it was effective at identifying hazards. Questions to be asked include:

- Was the team composition correct?
- Was the most appropriate PHA method used?
- Was the preparatory work complete?
- Were there any problems with the information provided to the team, particularly the P&ID's?
- Did the analysis move at the right speed?
- Did the team members feel that they had time to think around issues before the leader moved on?
- Were the team members bored?
- Does the report fairly and accurately summarize what was said?
- Was the report readable?

HAZARD IDENTIFICATION METHODS

TYPES OF HAZARD IDENTIFICATION

Knowlton points out that there are basically four methods of hazard identification:

- Experience based
- Augmented experience
- Analytical methods
- Creative methods

EXPERIENCE BASED

The first method for identifying hazards is that of individual experience and knowledge. Based on his or her experience with previous processes, a person can see analogous situations in the process being studied, and develop hazard scenarios from them. This type of experience can also be a team effort. Experience is also used for analyzing a new plant. A design engineer will know which standards to apply, and what has worked and not worked in the past.

AUGMENTED EXPERIENCE

The second type of hazard identification calls on the experience of groups of people outside the facility. For example, the collective experience of many engineers and operators may be used to develop a generic checklist that can be used in a wide variety of situations. The Failure Modes And Effects (FMEA) method uses augmented experience.

One way in which PHA teams commonly augment their experience is to call the vendor for the item of equipment that is being discussed on a speaker phone so that he can join in the discussion at that point.

ANALYTICAL METHODS

Analytical methods can use Boolean or stochastic logic (these terms are discussed in Chapter 16.) Analytical methods can be very useful because they apply quantitative analysis to situations where previously there had been little more than hunches and feelings. Their formality and rigor also gets over some of the problems of subjectivity and emotion that can occur with other methods.

CREATIVE METHODS

Finally, hazards can be identified with creative methods. Of these the most widely used is the HAZOP technique because it encourages an atmosphere for brainstorming within a structured discussion process. A less structured creative method is the What-If method.

TEAM APPROACH

Whichever of the above techniques is used, the basic idea is that a team of people gets together and systematically tries to find out how an incident could occur. This requires a different mind-set from what is normally needed to run a plant. Consequently many of the people who work in operations have trouble getting up to speed with it at first (particularly with regard to the guidewords "reverse flow", as discussed on page 156.)

As with any team effort, the quality of the leadership is critical. The leader must somehow allow people to dream up potentially bizarre situations in anticipation that one or two of these situations will actually turn out to be plausible. At the same time, he or she has to keep the PHA on track in terms of schedule and budget. Also, some team members enjoy the brainstorming process so much that they become counter-productive in terms of overall team effort. The leader needs to keep them focused, and to make sure that they move on to the next topic.

THE LEADER

The Team Leader is responsible for the overall conduct of the PHA, the quality of the final report, and for ensuring that the project is completed on time and within budget. Some organizations describe the person filling this role as a facilitator or coordinator; this is misleading. The leader's job consists of much more than reeling off a set of guidewords in sequence — he or she should actively lead the team. For example, if the team is discussing the handling of a highly hazardous chemical, the leader may decide to let the discussion run on for a long period of time during which some previously unthought of hazard may be identified. On the other hand, if the team is analyzing some low hazard situation, the leader should move the discussion along promptly.

The success of the PHA depends almost entirely on the quality of the team. That is why it is so critical to have the most experienced and knowledgeable people on the team, in spite of the fact that they are needed in so many other areas. Therefore, one of the most important of the leader's jobs is to ensure that he or she has a first-class team.

All team leaders have to work with the constraints of budget, schedule and the availability of key people. On the other hand, they have to meet their personal and corporate obligations to conduct a professional and responsible analysis. The leader must have the knowledge and experience to strike the right balance between these contending pressures.

The team leader has to strike a balance in the following areas:

- Creativity / Time-Wasting
- Openness / Rational Criticism Of Ideas
- Freethinking / Enforcing The Rules Of The Meeting
- A Thorough Search For Hazards / Meeting The Schedule And Budget

The leader should be from outside the immediate organization of the facility being analyzed. He or she does not necessarily have to be from a totally different organization, but must be free to report on hazards without fear of retribution. Moreover, an outsider is likely to be more successful at handling the "we've always done it that way, and it's never been a problem" issue. One of the leader's most important roles is to challenge commonly accepted practices.

The leader should have in-depth knowledge of how the process industries work, but will not know that particular unit in detail. Nor is it likely that he or she will be expert on the technology being discussed. However, the leader will have experience of other plants and other industries, and will share that experience with the PHA team.

One of the leader's most important roles is to generate imaginative discussions and to stimulate what is sometimes referred to as "helicoptic" thinking, i.e. the ability to rise up above a problem, and to see the forest as well as the trees. The leader has to get the team to "think the unthinkable." One way of doing this is by asking "stupid" questions, knowing that they will, at times, make him or her look a little foolish. Nevertheless, some of these "stupid" questions may generate a useful discussion, which in turn will encourage other people to think more broadly.

The leader should be sensitive to "throw-away" comments because these can lead to the identification of a hazard. For example, if the team member says that a certain valve "works OK now" the team leader might pick on the word "now" and find out what the source of the original problem was, and whether it could occur again. Similarly, the leader should be sensitive to jokes because they are often indicative of a deep-seated or endemic problem; it is the truth that lies behind a joke that makes it funny.

The leader needs to watch out for team members who dominate the discussion, and also for team members who hardly participate. It is important to make sure that everyone contributes their thoughts and experience, and that there is a balance between the members.

THE SCRIBE

The scribe records the discussion in an organized manner, usually using special software on a PC. Many teams choose to have the scribe's work displayed live on a screen so that the notes can be reviewed as they are being written, and so that the framework for the discussion is apparent to all. If this technique is used, and if the scribe is a bad typist, it is essential that he or she have a thick skin.

Sometimes the scribe is a full member of the team, and is expected to participate in the discussions. At other times, the scribe may be a secretary who is very good at recording notes, but who does not possess process or plant expertise. Either way, the scribe has to have the ability to screen the conversation and only to record what is pertinent. He or she is also usually responsible for preparing interim and final reports under the guidance of the team leader.

In addition to taking notes and preparing reports, the scribe may also have the broader responsibility of filing all of the PHA's. Questions that have to be answered in this context include:

- How are the PHA records to be managed?
- How are the recommendations and action items to be managed?
- How are the recommendations to be communicated?
- What media are to be used for storing the PHA records?
- How and when are they to be purged?
- Who has access to the PHA records?
- Who can modify the PHA records?

Finally, many teams use the scribe to organize the logistics of the PHA, including tasks such as making sure the P&ID's are up to date and available, and ordering rooms and equipment for the meetings.

OPERATIONS EXPERT

For plants that are already operating, the operations expert is a very important team member. Usually, this person will be a senior operator or supervisor. He will know the unit in great detail, and will be able to explain "how things really work around here." He will also have an excellent knowledge of past incidents, including those that were not necessarily serious enough to be reported as a part of the Incident Investigation process, but which nevertheless can provide very useful pointers to the PHA team.

The immense experience of the operations expert can, in some PHA's, have the drawback that he or she may have trouble thinking broadly. They know the unit so well, that they cannot visualize any other type of operation other than what they have seen.

PROCESS EXPERT

The process expert supplies knowledge about the process itself. He or she will often be a chemical engineer. Generally, they are expected to explain how the process works, what the basic chemistry is, and what would be the process impact of identified hazards. Sometimes the process expect is also the team leader. If this is the case, he or she must make it clear which role they are filling as the discussion progresses.

MECHANICAL / INSTRUMENT EXPERT

There should be someone on the team who represents the mechanical and/or instrument departments. Along with the operations expert, this person may also be able to represent the interests of the maintenance department.

SPECIALISTS

At times, it is appropriate to call in specialists for limited periods of time. For example, if the team is struggling with some specialized issues to do with corrosion, it may choose to call in a corrosion expert for that part of the PHA. Sometimes a team will save up its questions to do with a specialized area — say the shipping of samples to the lab — and deal with them all at once when that expert is present.

TEAM SIZE

The team size will vary according to the nature of the process, the analysis technique and the degree of experience of the individual team members. As shown, in Figure 5-1, the number of hazards identified will increase as the team size increases up to a certain point. Beyond that point, it is likely that adding more people will degrade the quality of the analysis because the discussion is more likely to be difficult to follow and control.

Figure 5-1

Effect Of Increased Team Size

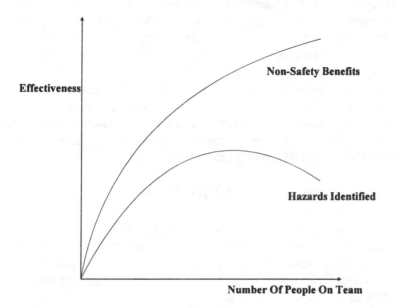

Figure 5-1 shows, however, that the non-safety benefits of a PHA continue to increase as the team size grows, even when the hazard identification effectiveness is declining. This is because the team members continue to learn about the process and the technology as more people join the team.

ALTERNATIVE TEAM APPROACHES

In order to reduce the cost of the PHA's, and to reduce the amount of time that they take, some companies have just two people conduct the PHA. They are the team leader and a process or operations expert. These two people work through all the P&ID's and develop a report that lists all their findings. This report is submitted to a committee which is drawn from both people who would have been on a conventional PHA team, and also from management and appropriate specialists. This committee then critiques the report that the two-person team has prepared.

This approach does reduce the number of man-hours consumed by the PHA, and may actually improve the effectiveness of the analysis because less time will be wasted; each person will be fully occupied making his or her full contribution. However, there are also a number of concerns. The first, and most serious, is that it reduces the potential for Employee Participation. PHA's are normally an excellent way of involving a wide range of employees in the process safety program, and exposing them to its philosophy and approach. The reduced team-size approach would reduce or eliminate this type of involvement. The second concern is that the team is less likely to come up with those "off-the-wall" ideas that provide fresh insights into how accidents can occur, and what can be done to reduce them. Finally, this approach may not be accepted by a regulatory agency.

BATCH PROCESSES

PHA methodologies were developed initially for large, continuous processes such as petrochemical plants and refineries. However, many plants are much smaller and operate primarily in a batch mode. Examples of such industries include: pharmaceuticals, terminal operations and food processing.

The classic PHA methods can be used for plants such as these, but some modifications may need to be made, particularly with respect to the nature of batch operations. Because time is a variable on these processes, an analysis of their operation is more complex than it is for a steady-state process. One way of handling this is to systematically work through the operating procedures — asking deviation guidewords at each step. For example, if the instruction is "Add 100 liters of water to V-100", the team might ask questions such as:

1. What if the vessel is over-filled? (High level)
2. What if the liquid is not water? (Contamination)
3. What if there is less than 100 liters of water available? (Low Flow)

Other "step" questions include:

1. Step done early
2. Step done late
3. Step omitted

Once the discussion for this step is complete, the team can then analyze the next step in the operating instructions.

PHA METHODOLOGIES

There is a wide variety of PHA methods. The methods listed by OSHA are:

- What-If
- Checklist
- What-If / Checklist
- Hazard And Operability Study (HAZOP)
- Failure Modes And Effects Analysis (FMEA)
- Fault Tree Analysis
- Other Appropriate Methods

No one of these methods is inherently any better than any of the others. They all have their place. Also, they are also often used in combination. For example, if the team has started an analysis with the HAZOP method (generally the most rigorous and time-consuming) it may gradually start to make increasing use of the What-If technique, particularly if it is found that many of the scenarios being discussed toward the end of the PHA are similar to what was discussed in depth at the beginning.

Actually, the differences between the methods are not as great as might appear at first, particularly when the PHA team is very experienced. For example, if the team is conducting one of the less-structured methods, such as What-If or Checklist, it will often be found that the team members are almost unconsciously working through the systematic HAZOP guidewords, even though their thoughts are not being verbalized or written down. Moreover, the methods often complement one another, particularly for a new project, when a series of PHA's of increasing depth, complexity and rigor can be performed as the plant design matures (*see* page 106.)

THE WHAT-IF METHOD

DESCRIPTION

The What-If [1] method is the least structured of the creative PHA techniques. It assumes that the team is composed of very experienced analysts who are able to identify incident scenarios based on their experience and knowledge. It is often a good method to use for Conceptual PHA's, where very little detail is yet available concerning the process, or the equipment that is being used. It allows the team to quickly focus on issues that are critical. There is no need to spend large amounts of time discussing guidewords that generate little or no significant insight. This also helps reduce the boredom of the meetings, a common problem with the more formal methods, such as HAZOP. What-If analyses are good at analyzing global issues, such as loss of utilities or the impact of a major fire.

Because it has relatively little structure, the success of a What-If analysis is highly dependent on the knowledge, thinking processes, experience and attitudes of the individual team members; the method does not structure a discussion in the way that HAZOP does. It does however allow the team members to be very creative — the lack of structure opens up their horizons. It also makes the method easy to learn and to use. Because there will be relatively little prompting from guidewords or line-by-line analyses, it is vital that the team members prepare very thoroughly before the group gets together; the free-ranging nature of the discussion will require that everyone be up to speed on the process and its general hazards before the meetings start.

The analysis itself is usually organized around sections of the P&ID, typically major equipment items, or operating systems (such as the condensing system of a distillation tower.) Team members ask questions such as "What-If there is high pressure?" or "What-If the

---

[1]   The phrase "What-If" is spelled here in the same way as it is printed in the OSHA regulation. It is hyphenated and the question mark is omitted, unless it is placed at the start of a longer sentence.

operator forgets to do this?" or "What-If there is an external fire in this area?"

Issues that can be discussed during a What-If review include the following:

- Electrical classification areas
- Emergency shut down systems
- Vents
- Flares
- Piping systems
- Truck/rail/ship/barge movements
- Effluents and drains
- Winterization
- Noise
- Leaks
- Machinery, including cranes, hoists and fork lifts
- Public access and perimeter fencing
- Adjacent facilities
- Buried cables
- Overhead cables
- Special weather problems, including freezing, fog, rain, snow, ice, high tides and high temperatures
- Toxicity of construction materials
- Demolition safety

WHAT-IF ANALYSIS OF DISTRIBUTED CONTROL SYSTEMS

Many plants are controlled by Distributed Control Systems (DCS). The analysis of these systems can be done with a customized What-If method. (Because, a PHA is a *process* analysis; consideration of electronic systems is out of scope. The discussion should consider only the process impact of the DCS.)

A DCS system consists of a set of cards or boards. Typically, about eight signals come from the plant into the board, as shown in Figure 5-2. The DCS circuitry processes the incoming signal and calculates one or more output signals which adjust control valve positions. Figure 5-2 shows a card that has eight input and eight output signals.

Figure 5-2

Sketch Of A DCS Card

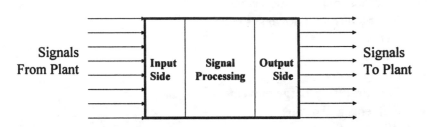

It can be seen from Figure 5-2 that there can be failure on the input side or on the output side. If there is a failure on the input side, two things will happen. First, the transmitted values to the DCS screen will disappear; the operator will have no knowledge as to what those values are unless someone goes outside to check. The second consequence of the failure of the input card is that the signals that it controls will fail, usually in their last position. This means that the control valves that are being controlled will "freeze" in place ("fail last position".) Therefore, if the operation was running smoothly before the input side of the board failed, there should be no sudden change in the operation. However, it is important for the PHA team to make sure that the output signals do fail in their present condition. If they cause the affected control valves to fail closed or open, the operation will be affected much more dramatically.

Failure of the output side of the DCS card is a more serious matter. This will generally cause all valves to fail in their safe position. This may sound acceptable, but it has to be remembered that the fail safe positions for the valves were selected assuming that there has been a total system failure. The failure of just one board will lead to just eight valves failing, with all the other valves on the plant in their normal operating position. This could create unusual and possibly unsafe situations.

For example, a total system shutdown may cause a pump to stop and its discharge valve close. In the case of a single card shutdown situation, however, it may be that the pump will continue to operate but that its discharge valve will close. This could cause the pump seal to leak hazardous chemicals. When evaluating such scenarios, there are three general rules to follow:

1. Sections of the process that remove energy should continue to operate. This includes cooling water systems and fin fans. Usually, the removal of energy from a process will move it toward a safer state.

2. Sections of the process that add energy should be shut down. This includes steam and electric heaters.

3. Sections of the process that contain large quantities of hazardous chemicals should be contained, i.e. the block valves around them should be closed, but the vent and drain valves may be opened, depending on the contents of the vessels.

The What-If DCS PHA, therefore, has two principal parts. First, it must evaluate the fail-safe condition of each valve, not only for a total system outage but also for a failure of the valves on the boards individually. Once this is done, the team should then evaluate the effect of a board failure on the rest of the system. One recommendation resulting from this analysis may be to move control loops from one board to another. In general, the components of a single system, such as a distillation column or a reactor, should be put together on one board.

WHAT-IF REPORTS

The lack of structure makes it difficult to record and to report a What-If analysis in a systematic manner. The final report will be discursive in nature. It will probably have the following sections:

1. Information Provided
2. Questions Raised
3. Discussion Around Those Questions

4. Actions Items
5. Recommendations

Table 5-2 provides an example of the framework for the final report.

Table 5-2

Example Of What-If Report Framework

What-If Question	Consequences	Risk	Action Item	Recommend -ation

THE CHECKLIST METHOD

The Checklist Method uses a set of pre-written questions to stimulate discussion and thinking. The questions are developed by experts who have conducted many PHA's and other similar analyses [2]. Checklists are not comprehensive — no PHA method can make that claim. Nevertheless, they should make sure that a complete range questions is asked, and that nothing that would be regarded as obvious is overlooked. Some areas for checklist questions are provided in Table 5-3.

---

[2] Detailed lists of What-If and Checklist questions are provided in the book Audit Protocols For Process Safety from Southwestern Books, scheduled for publication in 1998.

Table 5-3

Checklist Areas

I.   Loss Of Utilities
    A.   Steam (various pressure levels)
    B.   Cooling Water
    C.   Refrigerated Water
    D.   Process / Service Water
    E.   Instrument Air
    F.   Service Air
    G.   Boiler Feed Water
    H.   Nitrogen
    I.   Other Utility Gases
    J.   Fuel Gas
    K.   Natural Gas
    L.   Electrical Power
II.   Pressure Relief
    A.   Relief Valves
    B.   Rupture Disks
    C.   Flare Header
    D.   Flare
III.   Instruments And Controls
    A.   Local Instruments
    B.   Board Mounted Instruments
    C.   Distributed Control System (DCS)
    D.   Control Loops
    E.   Emergency Loops
IV.   Emergency Systems
    A.   Fire Water
    B.   Fire Fighting Equipment
    C.   External Fire
    D.   Runaway Reactions
V.   Human Factors
    A.   Operating Procedures
    B.   Training
VI.   Siting
VII.   Safety Systems

THE WHAT-IF/CHECKLIST METHOD

The What-If / Checklist method allows the team to think freely as if they are following a What-If approach, but also to provide some structure from the checklists. Typically, the team will first conduct a What-If, then follow it with a Checklist analysis.

HAZARD AND OPERABILITY METHOD (HAZOP)

DESCRIPTION

The HAZOP method is the most systematic and thorough type of PHA. Goodman points out that it is often used to ensure maximum compliance with the standard, even though its use may not be technically appropriate, and even though one of the other methods can be just as effective in a particular situation at finding the hazards in a particular situation. Nevertheless, the operating company's legal advisors prefer HAZOP because it is the most complete. In order to address this problem, he discusses the Focused What-If method, which has the thoroughness of the traditional HAZOP, but is not so time-consuming.

Because if its widespread use in the process industries, a variety of HAZOP approaches have been developed. They all work by dividing the unit into nodes or sections. A node represents a section of the process where a significant process change takes place. For example, a node might cover the transfer of material from one vessel to another through a pump. In this case the process change is the increase in pressure and flow that occurs across the node. Another node might include an overhead air-cooler on a distillation column. Hence, the process changes are change of phase of the process liquid and transfer of heat.

To save time, the leader and the scribe should select the nodes before the PHA starts. However, they do need to be flexible. If the team wishes to change the size of a node, they must be able to do so. Often, node sizes increase as the HAZOP progresses because many of the hazards have already been discussed.

Having selected a node, the team defines its purpose, and then systematically discusses deviations from Design or Operating Intent. If a plausible scenario is identified, the team then considers what the consequences could be (safety, environmental, economic) and makes a first estimate as to its likelihood. If the identified hazard is serious enough, the team issues a Recommendation or an Action Item.

The team examines the process functions within a node and then ask a set of structured questions around it. In the case of a flow of liquid, for example, the team will discuss the questions listed below. For each case, they will ask whether the scenario could happen, what it's consequences might be, and what is the likelihood of its occurrence.

1. More flow than intended
2. Less flow than intended
3. Reverse flow
4. No flow at all
5. Change in phase (for example, it may be that a certain fluid should be should be liquid, but could it be gas or solid?)
6. Change in viscosity

The power of the HAZOP technique becomes evident, even from this very brief overview. The plant is thoroughly analyzed because every line, every valve and every piece of equipment is incorporated into a node. Furthermore, all possible deviations within that node are explored.

Although the strength of the HAZOP method lies in its clear organization, it is important not to allow the analysis to become too rigid. If the team finds that it is talking about "reverse flow" even though the current guideword is "high flow", the leader should probably let the discussion continue. If he or she were to postpone the discussion until the "right" guideword, the current thinking and creativity may be lost. On the other hand, the leader must also keep the discussion focused on the issue at hand, and should control unstructured conversation. It is balancing forces such as these that makes the leader's job so challenging.

MECHANICS OF A HAZOP

General issues to do with running a PHA have already been discussed (*see* page 111.) It is important to make sure that as much work is done ahead of time as possible so as not to waste the time of the team members. One way of doing this is for the scribe and leader to predefine the nodes ahead of time, always recognizing that these may change once the PHA starts.

Whether it is done ahead of time or not, the scribe should enter the node description into the HAZOP software in detail. The start and stop points should all be identified. Generally, a node's boundaries will be block valves, although other piping items, such as a mixing tee, are often used. The scribe should also color in the node on the P&ID ahead of time, using a highlighter pen. A rigorous and precise definition of nodes is not usually of great benefit to the team discussion. However, it does provide the company with protection should the PHA be audited at a later date. It will show that no part of the plant was overlooked.

Most team leaders use colored highlighter-type pens to record the progress of the discussion and to define the boundaries of each node. Different colors are used so that the interfaces between the nodes are easily seen. At the end of the HAZOP, a quick glance at the P&ID's will confirm that all of the equipment and lines were discussed. The choice of color is not usually significant. However, some colors may have designated meanings. For example, the color blue may mean that the sections so highlighted were not discussed because they had been covered by another HAZOP, previous to this one. The color brown may designate items of equipment and piping that are deliberately being excluded from the HAZOP because they have been permanently disconnected from the main process. Yellow may indicate that a node has been defined but not yet discussed.

FAILURE MODES AND EFFECTS ANALYSIS (FMEA)

Failure Modes and Effects Analysis (FMEA) is a technique for determining the ways in which equipment items can fail, and what the consequences of such failures would be on the overall system reliability and safety. Traditionally, the method concentrates on individual equipment items, and often the components within those items. The method is not suitable for analyzing system effects, nor is it generally effective at considering human errors. Because FMEA is a detailed equipment analysis, it can only be used either on equipment that is already installed and operating, or on new designs that are almost complete.

Traditionally, the FMEA method has been widely used in the aerospace and nuclear power industries, but much less by the process industries. FMEA is most useful when combined with a method such as Fault Tree Analysis (FTA) that examines overall system effects.

As its name suggests, FMEA works by determining the failure modes for the item in question, i.e. the ways in which it could fail, and the consequences of each type of failure. Many items will have different types of failure mode, each of which can led to a different type of effect or consequence. Frequently, a failure may be partial, i.e. the item will operate in a degraded manner. In general, it is important to remember that most equipment items will fail in different ways and/or have partial failure modes. Therefore, once the PHA team has identified one failure mode, it is important not to move on right away to the next equipment item. Instead, different failure modes for that same item should be investigated.

Like other types of PHA, an FMEA should be carried out by a team. However, it is useful if one person can be assigned the task of researching information. For each item of equipment, the following information can be tabulated:

- The failure
- The effect of the failure
- Historical evidence to confirm the above two items
- The Mean Time Between Failures (MTBF). Frequently, this information is not available and is not really needed.
- Recommendations or findings

Table 5-4 provides some failure modes that can be considered for each equipment item (these items could also be used to develop a PHA Checklist.)

Table 5-4

Typical Failure Modes

I.   Automated Valves
   A.   Fail open
   B.   Fail closed
   C.   Partial Failure
   D.   Fail present position
   E.   Cyclic operation
   F.   Stem leaks
   G.   Material in body cavity
   H.   False position indication
II.   Pumps
   A.   Fail to operate
   B.   Fail to stop
   C.   Partial failure
   D.   Wrong impeller size
   E.   Cavitation
   F.   Seal leaks
   G.   Run backwards
   H.   Reverse flow through them
   I.   Deadheaded
   J.   Starved suction
   K.   Plugged strainers
   L.   Solids
   M.   Water present (often after turnaround)
III. Heat Exchangers
   A.   No flow on shell side
   B.   No flow on tube side
   C.   Partial flow
   D.   Fouling shell side
   E.   Fouling tube side
   F.   Fouling tube sheet
   G.   Leak shell to tubes
   H.   Leak tubes to shell

I.   Overpressure
IV.   Temperature Measurement
- A. Wrong material
- B. Badly welded thermowell
- C. Incorrect calibration
- D. Fouled thermowell
- E. Element partially out of thermowell
- F. Air leakage into thermowell
- G. Bad transmitter
- H. Bad setpoint
- I. Bad controller
- J. Incorrect alarm / interlock values
- K. Manual operation

V.   Flow Measurement
- A. Wrong material
- B. Wrong calibration
- C. Plugged orifice taps
- D. Wrong-size orifice plate
- E. Orifice backwards
- F. Corroded / Eroded orifice plate
- G. Transmitter error
- H. Wrong scale
- I. Wrong setpoint
- J. Drift
- K. Manual operation

VI.   Alarms
- A. Wrong setpoint
- B. Faulty reading
- C. Operator out of sight / earshot
- D. Alarm silenced / bypassed

VII.   Level Controls
- A. Stuck float
- B. Missing float
- C. Wrong calibration
- D. Broken sightglass
- E. Wrong setpoint
- F. Plugged taps
- G. Wrong seal leg fluid
- H. Seal leg fluid in process
- I. Freezing

VIII. Pressure Controls
    A. Plugged tap
    B. Wrong setpoint
    C. Corrosion
    D. Wrong calibration
IX. Weight Sensors
    A. Wrong calibration
    B. Drift
    C. Wrong scale (English / Metric)
X. Distillation Towers
A. Tray damage
B. Tray pluggage
    C. Missing trays
    D. Missing internal manways
    E. Flooding
    F. Downcomer damage / plugging
    G. Loss of reflux
    H. Corrosion
    I. Leaks in
    J. Leaks out
    K. Overpressure
    L. Loss of condensing
    M. Loss of feed
    N. Polymerization
XI. Vessels
    A. Leaks
    B. Overflow
    C. Overpressure
    D. Vacuum
    E. Flammability of vapor
    F. Loss of purge
XII. Compressors
    A. Deadheading
    B. Surging
    C. Solids in feed
    D. Leaks
    E. Lube oil leaks in
    F. Lube oil leaks out
    G. Overpressure
XIII. Reactors

A.   Runaway reaction
B.   Overpressure
C.   Loss of agitation
D.   Overtemperature
E.   Overfilling
F.   Catalyst high
G.   Catalyst low

Table 5-5 is an example of an FMEA form.

Table 5-5

Representative FMEA Form

Item Name and Number:

Comp-onent	Failure Mode	Caus-es(s)	Cons eque nce	Like-lihood	Risk	Action Item	Recomm-endation

FAULT TREE ANALYSIS

DESCRIPTION

The final PHA method identified by OSHA is Fault Tree Analysis
(FTA.) Unlike the previous methods, FTA can be used to develop
both qualitative and quantitative results. This section discusses just the
qualitative approach; quantitative analysis is described in Chapter 16.

A fault tree is a logic diagram that shows the combination of
events that have to take place before an accident can occur. The
method differs from the other PHA techniques discussed to this point
in that it is often more suitable for a single individual, rather than a
team. The logic associated with fault trees, particularly when they
grow to more than about ten base events, makes it very difficult for a
team to discuss.

TOP EVENT

Fault trees are looking for ways in which systems can fail; therefore, they start with the creation of a "top event", which represents a system failure. For example, consider the very simple operation shown in Figure 5-3. The operator opens the fill valve in the fill line. When the level, as read on the level gauge reaches 100%, he closes the fill valve, and then opens the valve in the drain line. When the level drops to 0% on the gauge, he closes the drain valve, and restarts the whole process.

<u>Figure 5-3</u>

<u>Over-Filling A Tank</u>

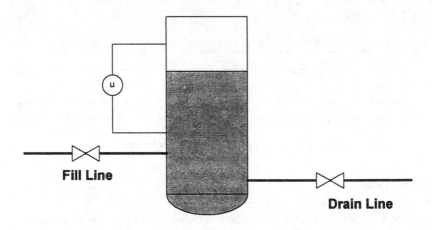

The Top Event for this situation is "Tank Overflows." This is the undesirable event to be avoided. This is shown in Figure 5-4, which is the first part of the fault tree.

Figure 5-4

Top Event

Tank Overflows

OR GATE

The first review of this system shows that there two possible ways in which this accident could happen. Either the level gauge fails OR the operator fails to notice that the level is at 100%. (Note that the word OR has been put into small capitals. This is to signify its use as a logical operator.)

Figure 5-5

First Level OR Gate

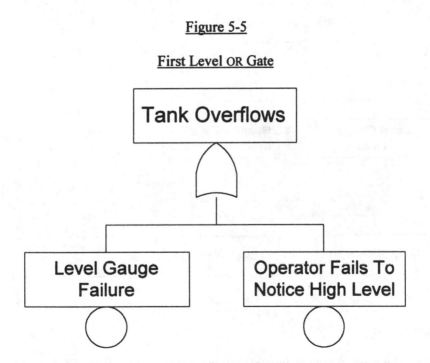

Figure 5-5 shows that, below the Top Event is an OR gate. Below it are two Base Events: 'Level Gauge Failure' and 'Operator Fails To

Notice High Level.' The Base Events are identified with a circle beneath its description box.

The base events can then be converted into intermediate events, and developed further. For example, there might be three reasons for the operator failing to notice high level. The first would be that he is busy elsewhere on the unit at the time that this tank fills up; the second may be that he is not paying attention; the third may be he inadvertently reads the wrong gauge. The tree that develops from this is shown in Figure 5-6. The original Base Event circle has now been replaced by an OR gate.

Figure 5-6

Second Level OR Gate

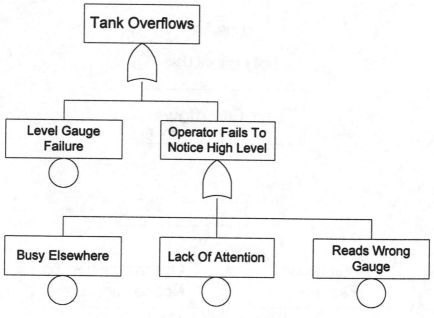

AND GATE

The second type of logical gate is the AND Gate. It requires all the events entering it to occur before system failure occurs.

In the simple example being used here, it might be that the level gauge frequently plugs up with a process chemical. Normally, the instrument mechanic is on call; he can come and clean out the instrument. However, if the mechanic is not available, the tree develops as shown in Figure 5-7.

<u>Figure 5-7</u>

<u>AND Gate</u>

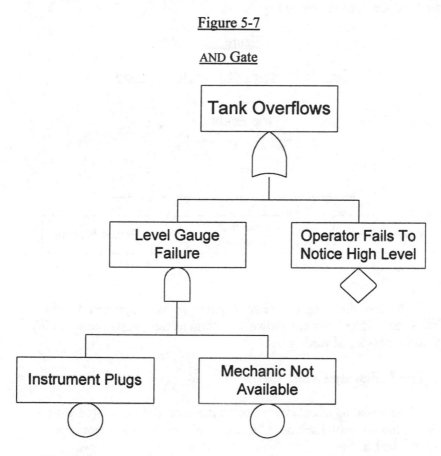

Note that the circle below 'Level Gauge Failure' has been replaced with the symbol which represents the AND Gate. Also, to keep the drawing as simple as possible, the OR Gate that had been developed previously has been replaced with a diamond, which represents an undeveloped event, the details of which have been excluded from the drawing in order to keep it as simple as possible.

FAULT TREE DEVELOPMENT

Although each fault tree analysis is unique, there are certain factors that tend to be common to them. For example, the Top Event can often be drawn with three Intermediate Events entering it through an AND Gate, as shown in Figure 5-8.

Figure 5-8

Generic Safety Fault Tree Development

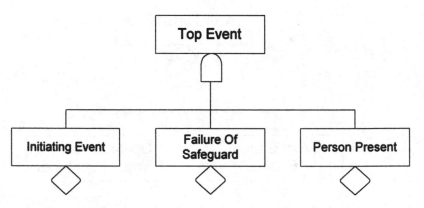

This tree shows that there are often three basic systems that go into such a tree. They are shown as Intermediate Events because they must be developed further.

1. The Initiating Event

The initiating event is what turns the potential associated with a hazard into an actual accident. It can be either a human error or a mechanical failure.

2. Failure Of The Safeguard(s)

A safeguard is a device or administrative activity that is specifically put into place to reduce either the consequences of an accident, or the probability of it happening (*see* page 116). Normal control systems and administrative systems (such as operating procedures and training) are not safeguards, although they are, of course, critically important to safe plant operations.

Examples of safeguard failures include:

- Failure of a relief valve due to pluggage
- Failure to train an operator in a special emergency operation
- Failure of a high level alarm
- Failure of a deluge system to operate.

3. Persons Present

If a person is not present when an incident occurs, then no one will be hurt. There was the potential for hurting someone, and there are likely to economic and environmental consequences. But if no one is present, no one can be hurt. As plants become increasingly automated, this factor becomes more and more important. On some remote gas plants, operator presence may be limited to just a few minutes a day.

On the other hand, many accidents occur during start-up and shutdown. It is at these times there are likely to be a lot of maintenance and construction people in the area. Therefore, it may be necessary to consider the possibility that many more people than normal will be in the area at the time of the accident.

DISCUSSION OF THE FAULT TREE METHOD

Fault Tree Analysis has the following strengths:

- It provides a good, graphical picture of the hazardous situation being analyzed
- Its logical approach is appealing to engineers and other technical people
- It rationalizes feelings and hunches
- It can be both qualitative and quantitative
- It is very useful in Incident Investigation

The method also has some weaknesses:

- It can be time-consuming and expensive
- The results depend heavily on the background and skills of the analyst
- It gives an appearance of completion — yet this may be misleading.

The last point is particularly important. Because fault trees are so logical, there is a tendency to believe that they must be complete. This is not the case; there will always be items that were overlooked. For this reason, some practitioners insert an event called "All Overlooked Items" directly below the Top Event. The only purpose of this is to remind them that there are always additional factors to take into account.

There are many texts that describe fault tree methodology in greater detail. For example, Sutton 1: 93 and Vesely et al.

DISCUSSION OF HAZARD IDENTIFICATION METHODS

Because PHA's have become so widely used, it is useful to examine some of the strengths and weaknesses of this approach to increasing system safety.

STRENGTHS

Availability Of Time To Think

One of the greatest benefits of a PHA is that it gives the selected experts time to think through the hazards associated with their units in a systematic and thorough manner. Such people are normally very busy with their day-to-day work, and rarely have chance to take an extended period of time to think deeply in a systematic and organized manner about the safety of the units for which they are responsible. Similarly, designers of new facilities are usually under considerable short-term pressures to get their work completed and sent on to construction. The PHA provides them with an opportunity to back up and to reflect on the overall safety of the facility that they have just designed.

This is one reason why the team leader should make sure that there are no distractions in the form of telephone calls or radio messages. He or she must ensure that this commitment of time is not violated by the demands of "real work." The team members are generally of high quality; they know the unit well, and they are highly respected. This means that there is constant pressure to take them off the PHA so that they can get on with their "real work." Resisting this pressure is one of the most important tasks of the team leader.

Cross-Discipline Thinking

One of the most valuables features of PHA is that it brings together people with different skills and backgrounds; this can lead to very fruitful brain-storming. It can also help flush out potentially hazardous assumptions. It could be that one department thought that another department was responsible for a certain activity, only to find that such is not the case.

Economic Payoff

Although the normal reason for carrying out a PHA is to improve safety, many companies feel that these analyses provide an economic payback. Unfortunately, such feelings are hard to verify. The problems with trying to determine economic payoff are those associated with all kinds of risk-benefit analysis. If a team identifies a high consequence hazard that has never actually occurred, and then recommends spending money to make its probability much lower, there is no direct financial benefit to the company. It is very difficult to justify spending funds on protecting against what has never actually happened.

Training

PHA's are an excellent training process for those who are unfamiliar with the process being analyzed. Even people with years of experience are often surprised to find that they learn a considerable amount about their process, particularly with regard to the original design decisions.

Development Of Process Safety Information

An important side benefit of the PHA process is that it puts management's feet to the fire with respect to developing up-to-date Process Safety Information. In particular, it ensures that the time and effort is spent on making sure that the P&ID's do indeed reflect the "as-built" condition of the unit.

WEAKNESSES

False Confidence

Given the investment that management makes in the PHA process, they tend to expect that the PHA team will uncover all hazards. Management sometimes has trouble understanding that the team — no matter how well qualified and no matter how serious their commitment to the PHA — will miss potential hazards. If there is an accident in a unit after the PHA has been performed, and the team did not identify that particular scenario, some people will use this to "prove" that PHA's "don't work." The defense that risk management by its very nature deals only in probabilities is seen as being evasive.

The second line of defense — that PHA's are not intended primarily to find hazards but to create an awareness among all employees and contract workers on how to look for hazards all the time — is much too abstract, even if it is true.

Safeguards present another source of false confidence. It is extremely unlikely that a highly hazardous situation has never been considered at all, hence safeguards will be present. This can lead to a problem because those safeguards may not actually be working as expected. However, because they are not normally used, this type of covert failure may remain in place for many years unless there is a good testing and inspection system. Hence, overall system safety may not be as good as anticipated.

Some safeguard devices are routinely tested. For example, most plants will bench test the set pressures of their relief valves every two years, at least. However, even this program has two potential weaknesses. First, the performance of a relief valve depends not only on its set pressure, but also on its capacity, i.e. how quickly it can remove vapors once it is open. It not common for this to be checked, even though it is critically important. Second, if the relief valve fails between tests, there is no way of knowing this. (Some safeguards, particularly electronic shutdown systems can test themselves, but mechanical equipment cannot.)

Team Composition

The hazards identified, and the recommendations developed, will depend in part on the personalities and experiences of the team members. If a particular accident has actually happened, for example, there will be a tendency to assume that it has a high probability of occurrence (and vice versa.) Such may not be the case — everyone tends to be strongly influenced by their own experience rather than dispassionate analysis.

Difficulties With Reporting

PHA reports can be very difficult to read and understand. During the PHA, the team may have had an exciting and insightful discussion into a particular situation. However, unless the quality of the written notes is of the highest standard, it is often difficult for others (even the original team members) to re-create the discussion later and to understand the rationale behind some of the recommendations.

Qualitative / Circularity

Most PHA's are qualitative. Some quantitative analysis may be carried out during a PHA, but the data used are generally no more than estimates based on experience and feelings.

Given that PHA team members and leaders are often engineers, the qualitative analysis of the PHA can be frustrating because it goes against the nature of their education — which is quantitative and numerical. This may explain the attraction of the Fault Tree method of analysis, because it does introduce an element of numerical rigor into the analysis.

Related to the above point is the issue of circularity. Hazards are often defined in terms of themselves. For example, a team may be discussing the guide words "high temperature." Implicit in the discussion is often the following train of thought:

1. We will discuss the hazards that can arise from "high temperature."
2. "High temperature" is defined as that temperature which can create hazards.

The only way of breaking this circle is to define the word "high" quantitatively. Then the phrase "high temperature" is replaced by a specific number such as 350°F. This is the reason why it is so important to define Safe Upper and Lower Limits for key variables before the discussion starts.

Abstraction

Given that PHA's work largely with low probability accident scenarios, they can be perceived as being very abstract. The great majority of the recommendations only show their benefit by preventing something from happened that has never happened anyway, and probably never will happen. This is not a weakness of the PHA concept, but it does constitute a major communication problem.

Boredom

Process Hazards Analyses can be long-winded and boring. It is difficult for anyone to maintain concentration and enthusiasm when the meetings drag on for days and weeks. This is a particularly serious problem for experienced PHA team members. They feel that they have seen it all before, and that they really do not need to go through a full guideword discussion of every point. To a large extent, they are correct in this opinion. The catch is that they may overlook an unusual situation that falls outside their previous experience.

Boredom also induces a pressure to "get on with it." (This is particularly true in afternoon meetings.) If one of the purposes of a PHA is to make sure that everyone — including those with a high level of experience — is forced to think the unthinkable, this feeling of urgency obviates that intent.

One response to the problem of boredom is to switch from one of the more rigorous methods, such as HAZOP, to a Checklist approach. This is probably where the judgment of an experienced PHA facilitator can be very valuable. He or she can decide when the team has enough experience to use one of the quicker methods without losing thoroughness or creativity.

Equipment Orientation

Although the OSHA standard explicitly includes human factors in the PHA element, the reality is that most PHA teams are composed of persons who have a technical background. As such, they tend to view the plant in terms of equipment rather than people. There is nothing wrong with this, but it is limiting.

For example, an equipment-oriented team might say, "The tank overflowed because the level controller failed." A people-oriented team may say, "The tank overflowed because the instrument mechanic did not service the level controller." A management-oriented team would say, "The tank overflowed because we did not have a good enough training program for out instrument technicians."

Single-Point Failures

Another potential problem with PHA's is that they are much stronger at identifying single-point failures than system problems. For example, if there is a control valve in a line within the node being discussed, it is almost certain that the impact of failure of the air supply to that valve will be discussed because it will come up at the guidewords "High Flow" or "Low Flow." However, in most cases, it is more likely that an instrument air failure will be system-wide. In this case a discussion as to the behavior of this particular valve is of limited value unless there is also a discussion of what else is occurring on the plant.

One way of addressing this problem is to have an analyst investigate complex system failures off-line, away from the PHA meeting. (This work can be done by the process engineer in many cases.) The team can then review the results of this analysis.

HIGH RISK SCENARIOS

Regardless of the method chosen, certain situations are often safety-critical. Some of these are described below.

REVERSE FLOW

Experienced operators often have trouble visualizing the concept of "reverse flow." They can visualize "high" and "low" flow because they have probably witnessed these, but reverse flow may be totally outside their experience. This is an important issue because reverse flow can be very serious because it could lead to the inadvertent mixing of highly incompatible chemicals.

REVERSE FLOW TO A UTILITY HEADER

The special reverse flow scenario: "Reverse flow to a utility header" is illustrated in Figure 5-9, which shows two lines, one containing a hazardous chemical and the other containing a utility such as nitrogen, steam or service air. There is a flow from the utility header to the process through a check valve that should prevent reverse flow from taking place.

Figure 5-9

Reverse Flow To A Utility Header Example

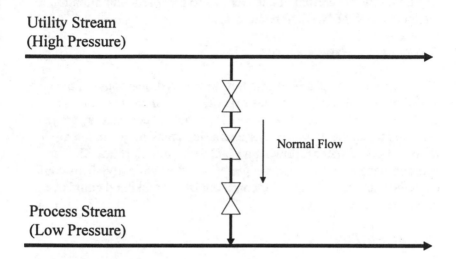

Utility Stream
(High Pressure)

Normal Flow

Process Stream
(Low Pressure)

The hazard scenario is as follows:

1. The pressure in the utility header falls (say due to a loss of the utility supply for a few minutes.)
2. The check valve fails .
3. The hazardous chemical flows into the utility header.
4. The hazardous chemical is rapidly distributed to many locations.

The probability of this incident is often quite high. Utility headers do lose pressure with some degree of frequency. They are also very effective at transporting materials around the facility. Check valves cannot be relied upon. Hence, this situation could occur. To make matters worse, this scenario is "memory-less", i.e. once the situation reverts to normal, there is no indication as to what happened.

Another problem with this scenario is that it can be very difficult to trace, once it has occurred. If the process chemical is used in many locations, it can take a long time to find out which of those locations is causing the leak.

The problem of cross-contamination is particularly serious when carrying out a HAZOP on the utility system itself. In principle, the discussion could be become very far-reaching and include almost the entire facility. Therefore, the leader has to pay particular attention to the location of the HAZOP boundaries.

WATER IN HOT HYDROCARBONS

Many process plants handle hot, heavy hydrocarbons. If they are mixed with cold water, a serious explosion could ensue. This is a particularly serious problem with storage tanks, because they are not designed to handle pressure. For example, a tank may contain a residual heel of water following a plant turnaround. If hot hydrocarbons enter the tank, the water will boil violently. This could cause hot liquids and vapors to come out of the tank, and could even cause the tank to rupture.

BLOCKED-IN PUMP

If a centrifugal pump is blocked in while it is still running, the liquid inventory within the pump will get hot due to the energy that is being added from the rotating impeller. Usually, the pump seal will leak and the pressure will be dissipated without creating a major hazard. However, it is possible that the liquid in the pump will boil, thus generating sufficient pressure to rupture the pump casing.

POLYMERIZATION

The inadvertent polymerization of a chemical in a vessel can cause an explosion, particularly if the system safety relies on the presence of an inhibitor that is mixed with the chemical. In this scenario, the chemical evaporates as heat is added (either from an external source or from a runaway reaction). The inhibitor remains in the liquid phase so the vapor that is formed does not contain any inhibitor. The vapor then condenses on the upper surfaces of the vessel — including the internals of the relief valve — and polymerizes. This has the effect of disabling a critically important safeguard.

EXTERNAL FIRE

One of the most common causes of vessel rupture is an external fire. The fire adds heat to the vessel, which heats up rapidly. Generally, relief valves are designed for this scenario. However, it is useful for the PHA team to check that there are no unusual external fire situations that might overload the relief valve and lead to a BLEVE caused either by a direct overload of the relief valve or by limitations in the flow due to the two-phase nature of the material being vented.

MULTIPLE USES OF EQUIPMENT

When equipment items are used for more than one purpose, there is a chance that a high consequence hazard may be created. For example, on one plant certain storage tanks were used to store a liquid that could freeze at ambient temperatures. Therefore, the tank contents were heated with internal steam coils. However, during turnarounds, the tanks were used to store another chemical. This second chemical was a monomer that could not freeze at normal temperatures and that could, if heated, polymerize rapidly and exothermically. Consequently, if during the plant re-start the operators omitted to clear the tanks of this second chemical before turning on the steam coils, the chemicals in the tanks could polymerize rapidly and exothermically, leading to an accident.

HUMAN FACTORS ANALYSIS

All PHA's are required by the OSHA standard to include an analysis of Human Factors. This presents the risk analyst and plant management with a dilemma. It is generally acknowledged that a large fraction, probably a majority of incidents, can be attributed to human error (of which operator error is just the final step in a series of design, implementation and organizational errors.) However human beings are impossible to 'model'. Each person is just that: a person. They come with all the trappings of personality that make them impossible to predict.

The intractable nature of human behavior at the individual level, in particular the near impossibility of modeling it quantitatively, means that risk analysts, many of whom have a technical background, usually feel more comfortable working with the equipment reliability because it can be predicted quantitatively. Therefore, when reviewing a risk management report of any kind, it is important to make sure that sufficient attention was paid to human error issues. This is also important from a regulatory point of view: in the PHA section of the PSM standard, OSHA specifically requires that human factors be addressed.

Related to Human Factors Analysis is the topic of Ergonomics. This subject examines the work place, and tries to find means whereby people can work more efficiently and more safely. Typical issues include the design of displays and seating.

Human Factors Analysis attempts to overcome the problems just described by analyzing human behavior and performance in a systematic manner. For example, human errors can usually be placed in one of the following categories:

- Initiating Error
- Error Following An Emergency
- Deliberate Error

INITIATING ERROR

There are three basic types of initiating error: Slips, Mistakes and Violations.

A *Slip* occurs when a person makes an error, even though that person knew what to do, and how to do it. It is formally defined here as follows:

A slip is a human error resulting from failure to carry out an intention, even though the person concerned had the capability, time, and equipment to successfully carry out that intention

Slips usually occur during normal, routine, non-stress situations. For example, an operator may routinely take two samples from a certain section of the plant every shift, and he may have successfully performed this action hundreds of times. Then, on one occasion, he *slips* up and inadvertently switches the samples.

A *Mistake* (sometimes referred to as a cognitive error) occurs when a person acts on an incorrect train of reasoning, often because he was not properly informed as to what to do or how to do it. Mistakes also occur in the following situations:

- Someone incorrectly follows a set of unfamiliar instructions
- A person is inadequately trained
- When an unusual situation occurs and a new sequence of actions has to be devised, often under considerable pressure, such as during an emergency shutdown.

A mistake is formally defined here as:

A mistake is a human error that is a failure in diagnosis, decision making, or planning

Mistakes can be further divided into errors that are "procedural" and errors that are "creative." A procedural mistake occurs when, for example, there is a lack of clarity in the operating instructions, thus causing the operator to misinterpret them. A creative mistake occurs when a brand-new situation presents itself, as sometimes occurs during an emergency, and the operator has to develop a response on the spot, often in a very short period of time.

Mistakes imply thinking, slips imply routine.

Violations are a special type of error. They occur when supervision or management knowingly decides to over-ride the normal operating or safety procedures. This is not necessarily wrong. Indeed, managers are paid to exercise judgment and responsibility in abnormal situations. Sometimes, however, the decision to violate a standard results in an incident.

A common example of a violation is when operations supervision chooses to ignore a lab result or an instrument reading, either because they do not believe it or because they are prepared to run the plant regardless of the information that they have received.

ERROR FOLLOWING AN EMERGENCY

A rule of thumb is that human error rates rise to 50% during an emergency, i.e. there is a 1 in 2 chance that a person will do the wrong thing in such a situation. Therefore, if an operator is called upon to perform say six tasks during an emergency, the chance of getting them all right is 0.5^6, i.e. 1.6%. Brandes points out that some people suffer from "panic induced impotence" during an emergency; they freeze and do nothing at all. The upshot of all this is that people should not be asked to make decisions during an emergency — they should carry out a few automatic actions in which they have been thoroughly trained, and then turn over responsibility to a qualified emergency response team.

INTENTIONAL ERROR

If a plant worker deliberately decides to create an incident, there is very little that can be done to prevent him from doing so. This problem is not one that can be addressed by process safety principles. Similarly, external sabotage is a topic that has to be handled by law enforcement authorities.

SITING

The OSHA standard requires that Facility Siting be considered. This can be handled with a checklist. For a large unit, it may be necessary to run the checklist for various types of siting, such as:

- General outside operations
- Control room
- Analyzer houses
- Warehouses

Topics that can be considered when analyzing siting include:

- Electrical classification areas.
- Maintaining a minimum distance (say 75') between vehicle unloading areas and storage areas (accidents often occur at loading racks, so it is important to make sure that they do not spread into process areas.)
- Ensuring that manually operated emergency isolation valves, fire hydrants and other emergency equipment are outside the processing zone so that they can be accessed during an emergency.
- Sloping ground away from storage areas.
- Isolation of administration buildings from process areas, and ensuring that administrative personnel can escape safely.
- Putting excess flow valves on tank unloading lines.

Additional help to do with siting can be obtained from API (American Petroleum Institute) Recommended Practice 752, and from Guidelines for Evaluating Process Plant Buildings for External Explosion and Fire (Center for Chemical Process Safety.)

CONCLUSIONS

Process Hazards Analyses are a fundamental part of any process safety program. Only when hazards have been identified can action be taken to address them. During the compliance phase of PSM, the HAZOP method was widely used. However, this method is very resource-intensive, and future PHA's may be conducted using What-If and Checklist. These can be just as effective at identifying hazards, and often take less time.

Chapter 6

OPERATING PROCEDURES

INTRODUCTION

Operating procedures are important because they describe the interface between the people who run the plant and the equipment for which they are responsible. Nevertheless, in spite of their importance, operating procedures are often found to be below par when a facility is audited. For example, one survey (Chemical Process Safety Report) noted that 22% of all PSM citations are for deficiencies in operating procedures. Moreover, Michael Marshall, an OSHA incident investigator, is quoted in the same report as saying that this is "far and away the most frequently cited element."

But the need for good procedures goes beyond regulatory compliance. They are a vital part of any smooth and safe operation. Many sources indicate that between 40 and 70% of accidents are caused by operating error (Chemical Engineering, October 1995.) High quality procedures, properly linked to the training program, will help reduce this accident rate. In addition, emergency operating procedures will help ensure that small incidents are properly contained before they become major accidents that threaten people and property.

This chapter provides an overview of the process for writing operating procedures. The same concepts are discussed in considerably greater depth in the companion volume in this series of books on process safety management: Writing Operating Procedures For Process Plants (Sutton: 2).

THE OSHA REGULATION

(1) The employer shall develop and implement written operating procedures that provide clear instructions for safely conducting activities involved in each covered process consistent with the process safety information and shall address at least the following elements.

(i) Steps for each operating phase:

(A) Initial startup;

(B) Normal operations;

(C) Temporary operations;

(D) Emergency shutdown including the conditions under which emergency shutdown is required, and the assignment of shutdown responsibility to qualified operators to ensure that emergency shutdown is executed in a safe and timely manner.

(E) Emergency Operations;

(F) Normal shutdown; and,

(G) Startup following a turnaround, or after an emergency shutdown.

(ii) Operating limits:

(A) Consequences of deviation; and

(B) Steps required to correct or avoid deviation.

(iii) Safety and health considerations:

(A) Properties of, and hazards presented by, the chemicals used in the process;

(B) Precautions necessary to prevent exposure, including engineering controls, administrative controls, and personal protective equipment;

(C) Control measures to be taken if physical contact or airborne exposure occurs;

(D) Quality control for raw materials and control of hazardous chemical inventory levels; and,

(E) Any special or unique hazards.

(iv) Safety systems and their functions.

(2) Operating procedures shall be readily accessible to employees who work in or maintain a process.

(3) The operating procedures shall be reviewed as often as necessary to assure that they reflect current operating practice, including changes that result from changes in process chemicals, technology, and equipment, and changes to facilities. The employer shall certify annually that these operating procedures are current and accurate.

(4) The employer shall develop and implement safe work practices to provide for the control of hazards during operations such as lockout/tagout; confined space entry; opening process equipment or piping; and control over entrance into a facility by maintenance, contractor, laboratory, or other support personnel. These safe work practices shall apply to employees and contractor employees.

OSHA GUIDANCE

Operating Procedures and Practices. Operating procedures describe tasks to be performed, data to be recorded, operating conditions to be maintained, samples to be collected, and safety and health precautions to be taken. The procedures need to be technically accurate, understandable to employees, and revised periodically to ensure that they reflect current operations. The process safety information package is to be used as a resource to better assure that the operating procedures and practices are consistent with the known hazards of the chemicals in the process and that the operating parameters are accurate. Operating procedures should be reviewed by engineering staff and operating personnel to ensure that they are accurate and provide practical instructions on how to actually carry out job duties safely.

Operating procedures will include specific instructions or details on what steps are to be taken or followed in carrying out the stated procedures. These operating instructions for each procedure should include the applicable safety precautions and should contain appropriate information on safety implications. For example, the operating procedures addressing operating parameters will contain operating instructions about pressure limits, temperature ranges, flow rates, what to do when an upset condition occurs, what alarms and instruments are pertinent if an upset condition occurs, and other subjects. Another example of using operating instructions to properly implement operating procedures is in starting up or shutting down the process. In these cases, different parameters will be required from those of normal operation. These operating instructions need to clearly indicate the distinctions between startup and normal operations such as the appropriate allowances for heating up a unit to reach the normal operating parameters. Also the operating instructions need to describe the proper method for increasing the temperature of the unit until the normal operating temperature parameters are achieved.

Computerized process control systems add complexity to operating instructions. These operating instructions need to describe the logic of the software as well as the relationship between the equipment and the control system; otherwise, it may not be apparent to the operator.

Operating procedures and instructions are important for training operating personnel. The operating procedures are often viewed as the standard operating practices (SOPs) for operations. Control room personnel and operating staff, in general, need to have a full understanding of operating procedures. If workers are not fluent in English then procedures and instructions need to be prepared in a second language understood by the workers. In addition, operating procedures need to be changed when there is a change in the process as a result of the management of change procedures. The consequences of operating procedure changes need to be fully evaluated and the information conveyed to the personnel. For example, mechanical changes to the process made by the maintenance department (like changing a valve from steel to brass or other subtle changes) need to be evaluated to determine if operating procedures and practices also need to be changed. All management of change actions must be coordinated and integrated with current operating procedures and operating personnel must be oriented to the changes in procedures before the change is made. When the process is shutdown in order to make a change, then the operating procedures must be updated before startup of the process.

Training in how to handle upset conditions must be accomplished as well as what operating personnel are to do in emergencies such as when a pump seal fails or a pipeline ruptures. Communication between operating personnel and workers performing work within the process area, such as nonroutine tasks, also must be maintained. The hazards of the tasks are to be conveyed to operating personnel in accordance with established procedures and to those performing the actual tasks. When the work is completed, operating personnel should be informed to provide closure on the job.

ANALYSIS OF THE REGULATION

The different sections of the standard are discussed in detail through the remainder of this chapter. Certain key concepts are worth emphasizing, however. One of these is that the operating procedures must be written down. In many facilities, the tasks that the operators have to perform are well understood. However, not all of this knowledge has been written down. Consequently, the procedures do not meet the regulation's requirements.

The regulation stresses the importance of emergency procedures, and what to do if there is an accident. This may be another area where actual performance is good, but there is insufficient written material to meet the requirements of an audit.

Operating procedures must be certified annually. This can be a demanding assignment because an experienced member of staff will have to check that what is written down reflects what is actually being done. Also, the procedures must incorporate the effects of any changes that have been made in the previous twelve months. This means that the person doing the certification has to review the Management of Change and Prestartup Safety Review systems to make sure that nothing has "slipped through."

ELEMENTS OF OPERATING PROCEDURES

DESIGN AND OPERATING INTENT

Before starting to write procedures, management and the operators must clearly define how the plant is to be run; in other words the target conditions for all flows, pressures and temperatures must be specified, along with the allowable deviations from those target conditions (*see* page 91.)

WRITTEN INSTRUCTIONS

Operating procedures must be written down in order to ensure that all of the necessary information has been communicated to the operators correctly, and so that they can be validated and audited. Sometimes, the operators may already have a good set of verbal procedures. In these situations, the procedures-writing project consists largely of recording the verbal information that already exists, checking it, and then writing it down in a clear and organized manner.

The process of writing the operating procedures can help determine the optimum way of operating the plant. For example, on a four-shift system, it is common for each shift to have its own way of running a unit. Each way may be quite acceptable and safe, and the differences between them may be minor. Nevertheless, it is likely that one of the four methods of operation will be superior to the others. The act of writing the procedures can generate a useful discussion as to which method to choose.

OPERATIONS

Operating procedures are written for the operators and their line managers. There are other users, such as engineers and auditors, but generally their requirements are secondary; the operators are the primary customers of the procedures-writing project. Therefore it is vital that the operators feel that they "own" the manual and that is reflects their knowledge, goals, concerns, opinions and needs. Ownership is always important — everyone likes to feel involved in what is going on, and that their opinions count for something. This is basic human nature. In the case of an operating manual, if the operators feel involved with its development and writing, they will be much more willing to use it than if it is merely handed to them as a finished product, written by someone else.

TERMINOLOGY

The vocabulary and terminology used with regard to operating procedures are not consistent from company to company. Because there is no industry standard, each company and facility will develop its owning meaning for particular words and terms in order to suit its particular operation. However, it would be useful to have some standard terminology for words and phrases such as "procedure" and "SOP". The terminology used in this chapter is explained below.

An *Operating Instruction* describes a single task. Operating instructions should consist of short, sharp sentences, written in the imperative tense, and starting with an action verb. Examples are:

- Measure the temperature
- Fill the truck
- Read the logbook

There should be no more than one instruction per sentence. Hence:

Open the valve, then stop the pump.

becomes

- Open the valve
- Stop the pump

An *Operating Procedure* consists of a set of operating instructions for completing a well-defined task with a clearly defined scope. Examples include the following:

- "How to start Pump, P-100"
- "How to check the bearings on Compressor, C-201"
- "How to organize samples to be taken to the lab"

Typically, an Operating Procedure should not be more than a page or two in length, and should contain between 10 and 20 instructions. If it is longer than this, it probably should be broken down into sets of smaller procedures. Figure 6-1 illustrates the concept of an Operating Procedure made up of Tasks or Operating Instructions.

Figure 6-1

Operating Procedure/Operating Instructions

```
┌─────────────────────────────────────┐
│  Operating Procedure                 │
│                                      │
│      Task 1                          │
│      Task 2                          │
│      Task 3                          │
│      . . .                           │
│      etc.                            │
│                                      │
└─────────────────────────────────────┘
```

An *Operating Manual* is a book that contains a complete set of operating procedures. Typical titles for Operating Manuals would be:

- Utilities Area — Emergency Response
- Tank Farm — Commissioning and Start-Up
- Crude Unit — Shut Down

A *Standard Operating Procedure (SOP)* is a generic procedure that can be used as the foundation for situation-specific operating procedures. For example, the SOP "How to start a centrifugal pump" can be used as the basis for the specific procedure: "How to start centrifugal pump, P-100."

The use of SOP's can greatly improve the quality of the final procedures, and can drastically reduce the development time for the complete manual. Quality is improved because the SOP can be very carefully reviewed by all the pertinent disciplines (operations, process, electrical, maintenance, etc.) to ensure that it represents the very best understanding of how to perform this particular task. Development time is reduced because the persons writing situation-specific procedures only have to edit a "go-by"; they do not have to start from scratch.

FIVE VOLUME STRUCTURE

One of the biggest practical problems with writing procedures is that they are usually intended to satisfy multiple objectives. It is important to clearly distinguish these objectives from one other, and to organize the manual appropriately. A simple example of the confusion of purpose that can spring from not clearly delineating such objectives is shown in Table 6-1.

Table 6-1

Example Of Confusion Of Objectives

PROCEDURE FOR FILLING THE WEIGH TANK	
Step 1	Start the Dilution Oil Pump, P-1205.
Step 2	Add ten (10) 50 lb. Bags of catalyst powder to the Weigh Tank, T-1230.
Step 3	Add 200 gallons of process water to T-1230 using batch meter FQ-1209. This will dissolve the catalyst powder.
Step 4	Measure the specific gravity of the solution. It should be in the range 1.105 - 1.115.
Step 5	Add an additional 100 gallons of water.

Strictly speaking, the second sentence in Step 3 — "This will dissolve the catalyst powder" — is redundant given that Table 6-1 should contain just operating *instructions*. The sentence is actually an item of process information, and should be located elsewhere in the manual. Although a single instance such as this is trivial, its repeated occurrence can seriously degrade the usability of the manual.

In order to minimize potential confusion of purpose, such as the above, it is suggested that the operating procedures be organized into five volumes or sections:

Volume I	—	*Operating Instructions*
Volume II	—	*Operating Information*
Volume III	—	*Emergency Procedures*
Volume IV	—	*Troubleshooting Manual*
Volume V	—	*Initial Start-Up Manual (where needed)*

These five volumes can be supplemented with a *Writer's Guide* and a *Project Manual*.

VOLUME I — OPERATING INSTRUCTIONS

Volume I provides instructions on how to actually operate the unit. It can be divided into the following sections:

1. Startup
2. Shutdown
3. Normal Operations
4. Temporary Operations
5. SOP's and Generic Procedures
6. Vendor Manuals

MODULAR STRUCTURE OF VOLUME I

On most plants, change is a constant. Almost every day there is a change to the equipment, the company organization or the operating targets. Most of these changes require that the operating procedures be updated because the procedures represent the interface between the plant and the people who are running it. Unfortunately, most operating procedures do not readily lend themselves to being updated because a small modification to a procedure in one place may lead to changes elsewhere in the manual. Frequently, it is difficult to locate the sections of the manuals that need to be changed. Also, the changes may create a need for a major re-formatting of the whole document. This is why keeping a manual up-to-date is typically more difficult than writing the first edition. Consequently, most companies do not update their operating procedures as frequently as they should. Therefore, after a few months the manual is out-of-date; after a year or two it may be so badly out-of-date as to be largely unusable or untrustworthy. At that point, it may be easier to start a brand-new procedures-writing project, rather than trying to update the old manual.

In order to avoid this waste of effort, it is very important to design the procedures for ease of maintenance and updating. Over the long-term, this is likely to be more important than expediting the production of the first edition of the manual.

Another business area which has struggled with these same issues is the software industry. Modern computer programs are extremely large, and the effects of even small changes in the programs can be both far-reaching and subtle. In response to this, software engineers make strict use of the concept of "publicly declared modules that are privately maintained." What this means is that all software code is divided into manageable modules. Depending on the computer language being used, these modules are called sub-routines, functions or procedures. Each module has an explicitly defined purpose and function, and, because its title is "publicly declared," it can be called by any other module.

Modules can only call one another through their title blocks. It is not possible to move from one module to the middle of another module. The advantage of this approach is that if a module has to be changed its internals can be updated and modified — "privately maintained" — without having any impact on the overall structure. As long as the title of a module is not changed, a writer is free to change the contents of a module without having to worry about system-wide implications.

The same principle can be applied to the Operating Procedures within the Operating Manual. Figure 6-2 shows the titles of a series of modules, which correspond to the definition of Operating Procedures provided above. These are executed in the order shown. The internal "code" for one of the modules — "Heat V-100" — is shown.

Figure 6-2

Procedures Modules

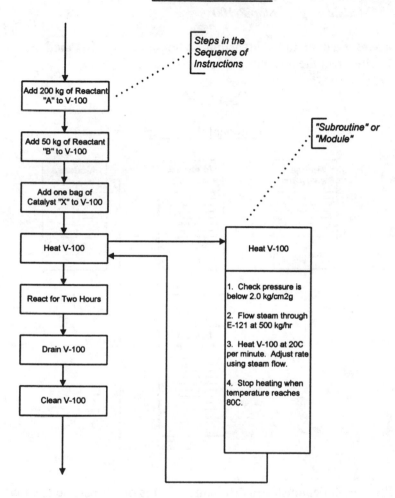

MODULE LINKAGE

The linkage of the modules may involve some branching based on conditional instructions, as seen in Figure 6-3, which shows a sequence of four activities in series. They are:

Module 1	*Start Pump P-100*
Module 2	*Fill V-100,* OR
Module 3	*Fill V-101*
Module 4	*Stop P-100*

There is a conditional branching statement in the first module, which will direct the operator either to V-100 or V-101.

Figure 6-3

Conditional Branching

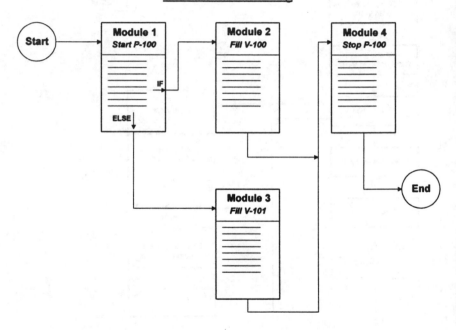

Figure 6-3 demonstrates that, although it is only possible to enter a module through its title block, it is possible to exit from the middle of it.

TOP-DOWN DEVELOPMENT

Figure 6-2 showed a fairly simple sequence of Operating Procedures, linked together to form a "macro" or higher level program. For very large systems, even this may be too detailed. In such cases, it is suggested that another software concept — that of "Top-Down Development" be considered. This is illustrated in Figure 6-4. In this example, there are five levels of increasing development.

- The first level is simply a single block — *The Operating Manual*.
- One level down the unit is divided into four major operating areas: Utilities and Units 100-300.
- At the third level, each of the operating areas is divided into five types of operation as shown (in Figure 6-3, this is shown for Unit 100 only.)
- Each of the operating areas is further divided into operating blocks. Block 3 — *Start Compressors* — is shown.
- At the fifth, and final level, are the actual modules or operating procedures. There are four of them for this particular block: Modules 3-1 to 3-4.

Figure 6-4

Top-Down Development

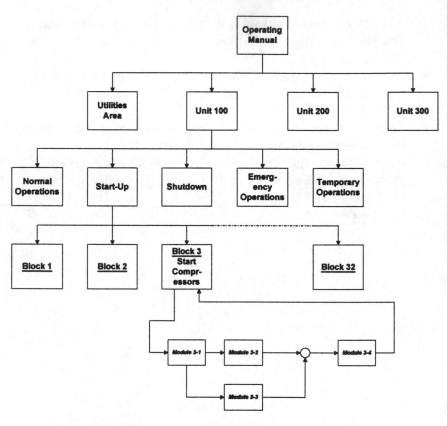

This approach is the basis of the aphorism:

Design the manual to be complete — design the manual to be incomplete

What this means is that, at the beginning of the procedures-writing project, the amount of detail is extremely limited, hence the manual is incomplete. However, the manual is complete because it contains within it all the operations that are to be carried out. As detail is added, the manual will always have a complete structure. Yet there will always be a need for more detail (often to reflect changes that have occurred in the process since the manual was first published.) These changes and additions can be made within the pre-defined framework.

MODULE DESIGN

Operating procedures should be written so as to be consistent with one another. Each module should have exactly the same appearance as all the other modules. There are two reasons for this.

1. A consistent style makes the operators' jobs easier. If the procedure modules are all the same design, the operators will find it much easier to locate the information that they need, even if the procedure itself is unfamiliar to the operator.

2. Consistency helps the writers. By pre-designing the format as much as possible, the writers are less prone to the problems associated with writer's block. Most people find it much easier to "fill in the blanks" than to write creatively on a blank piece of paper.

It is suggested that each module or Operating Procedure be divided into three principal sections:

1. The Title Block
2. The Operating Instructions
3. The Authorization Block

Figure 6-5 is an example of this structure.

Figure 6-5

Operating Procedure Structure

Title Block	
Operating Instructions	

	Action	Response/ Discussion
Step 1		
Step 2		
Step 3		
etc.		

Authorization Block	

THE TITLE BLOCK

The Title Block can contain the following elements. (In practice, it will not usually be necessary to include them all, but they are shown here for completeness.)

- Module Name
- Purpose of the Module
- Special Safety Considerations. These are supplemental to standard safety issues
- Discussion
- P&ID's and other Reference Documents
- Company/Plant
- Module Number
- Page Number/Date of Issue

One of the practical problems that procedures writers face is how much of this type of detail to include. It is tempting to include a large amount of information, particularly concerning safety issues. However, if the amount of information becomes so large as to make the actual operating procedures difficult to use, then the net effect on the operation may be deleterious.

THE OPERATING INSTRUCTIONS

The instructions can be laid out using a set of columns. As with the Title Block, in most cases it will not be necessary to use all of these columns.

1. The first column in the instructions is the *Step Number*. The order in which the instructions are listed is usually critical: they must be carried out in the sequence shown.
2. The second column identifies the *Person* who has to perform the action.
3. Next is the *Action* column that contains the actual instructions. Each instruction should be on a separate line.
4. The fourth column is for *Response/Discussion*. This provides information as to what the response should be to the action that was just taken. For example, if the instruction was to start a pump, this column can provide information regarding the anticipated discharge pressure and flow rates. If the observed values are different, the operator knows that corrective action is required.
5. The next column is *Revision Number*.
6. Finally, there is a column for a *Check Box*, that can be used by the operator to indicate that this particular action was performed successfully.

An example of this type of layout is provided in Figure 6-6.

Figure 6-6

Layout Of Operating Instructions

STEP	PERSON	ACTION	RESPONSE DISCUSSION	REV	√
1	B-OP	Check that the level in V-100 sightglass is at least 30%.	Do not start P-100 until the 30% level is reached.		
2	B-OP	**CLOSE THE BLOCK VALVE ON THE OUTLET OF P-100**		3	
3	A-OP	Place FRC-121 in manual	This may cause a small temperature swing in C-121.		
4	A-OP	Manually set FRC-121 to 25% open			
5	B-OP	Open the bypass valve around FCV-121 two turns			
6	B-OP	Start P-100	The discharge pressure should be 125 psig.		
7	B-OP	**OPEN THE BLOCK VALVE ON THE OUTLET OF P-100**		3	
8	A-OP	Put FRC-121 on automatic			
9	B-OP	Adjust the set point of FRC-121 for a flow of approximately 2000 lb/h			

THE AUTHORIZATION BLOCK

One of more managers must sign off on the procedure before it can be officially released. Their names and signatures go here, along with the signatures of the operator who provided the information and his supervisor. The signing of procedures is often a major bottleneck in a procedures-writing project. Chapter 17 provides a discussion concerning the sign-off process for process safety documentation, including operating procedures.

GRAPHICS

Where possible, procedures should be supplemented with graphics. Not only do graphics provide information and explanation, they also make the manual more attractive. This is important because a manual that is attractive is more likely to be used. The following types of graphics can be used.

- Photographs
- P&IDs
- Personalized Sketches
- Maps/Plot Plans
- Calculation Sheets (usually to go with batch operating procedures and/or checklists)
- Equipment Drawings
- Column Profiles

TWO PAGE MODULES

FACING PAGES

The modules should have a length of two pages because that is what the reader sees whenever he or she looks at an open manual. This issue is discussed by Weiss, who points out that all operating manuals are written in modular form, whether the writer realizes it or not; and the modules are called pages, of which the reader sees two at a time. Therefore, it makes sense to have the ideas be organized so as to fit the page, not the other way around. In other words, procedures should be presented in two-page chunks. In practice, it is not always appropriate to follow this rule strictly. Nevertheless, the idea of making the information fit the layout rather than the other way around is well worth considering.

One way of achieving the goal of two-page modularity is to have a page of text and a page of graphics facing one another. The text is put on the right-hand side because it is where most readers automatically look for written material (it is how single-page reports are laid out, for example.)

Placing text and graphics opposite one another in this fashion means that two sheets of paper are used. Therefore a complete module has four pages, as shown in Figure 6-7 (one sheet of paper has two pages on it.) The first page of the first sheet shows the title of the module. The second page of the first sheet has the graphic on it. The first page of the second sheet has the text. The second page of the second sheet (the fourth page of the complete module) can contain any overflow text.

Figure 6-7

Two-Sheet / Four-Page Module

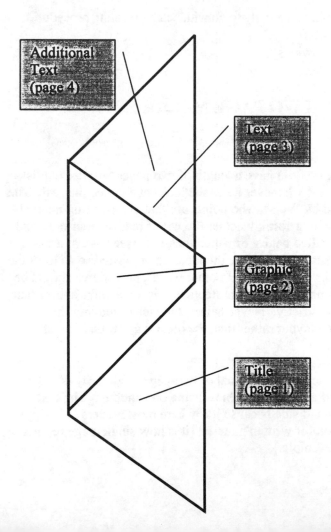

If a change is to be made to this particular procedure, all that has to be done is to pull out this two-sheet/four-page module, rewrite it, and reinsert it. *Nothing else changes.* This matches the concept of software sub-routines discussed earlier. The maintenance of operating procedures now requires much less effort than the normal process of removing pages, changing them, and then re-inserting them, all the time hoping that it will not throw the complete page-numbering system off.

MINIMALIST WRITING

The goal of minimalist writing is to present information in a manner appropriate to the way in which adults actually learn. There are various elements to minimalist writing, not all of which are useable, or even safe, were they to be applied to the process industries. Nevertheless, two of the ideas, the minimization of "soft" materials and the provision of a good index, are very applicable.

Many operating manuals are cocooned in a cloud of "soft" materials such as introductions, summaries, generic safety instructions and company mission statements. None of these materials really helps the operator do his or her work better or more safely; instead they pad the manual with material that gets in the way, thereby increasing the time it takes find useful information. In response to this, the minimalist approach eliminates as much as possible of these soft materials. In the case of operating procedures, the manual should tell the operator how to operate the unit — no more, no less. All other material should be located in other volumes of the operating manual, or else completely removed.

Furthermore, a minimalist approach condenses the style of writing. A person may *say:*

"Next, you need to go over to Pump P-100 and get it started."

The corresponding minimalist instruction would read:

"Start Pump P-100."

All redundant words and phrases have been removed. In principle, this sounds easy; actually, it is not. Engineers often use the passive tense and long sentences. So the instruction: "Having started P-100, start flow into V-100 using FCV-121, first making sure that the pressure in V-100 is not greater than 20 psig." becomes:

1. Start Pump, P-100.
2. Measure the pressure in V-100.
3. If the pressure is less than or equal to 20 psig, start flow from P-100 to V-100.
4. If the pressure in V-100 is greater than 20 psig, stop this procedure.

In this example, the minimalist approach actually uses more words and paper than the original sentence. Minimalist writing is not quite the same as short writing; it is a form of writing that reduces the number of words to the minimum needed.

VOLUME II — OPERATING INFORMATION

Volume II provides the operator with the information that he or she needs in order to do their job efficiently and safely; it does not contain any actual instructions. (Many of the Process Safety Information items in the OSHA Process Safety standard can be located in Volume II.) A representative Table of Contents for a Volume II is shown in Table 6-2.

Table 6-2

Representative Table of Contents for Volume II

I. **OPERATOR DUTIES AND RESPONSIBILITIES**

    A.   OUTSIDE OPERATOR
        1.   Shift Change
        2.   Visual Checks
        3.   Coordination with DCS Operator
    B.   INSIDE OPERATOR
    C.   UTILITY OPERATOR
    D.   TASKS LISTS

II. **DESCRIPTION OF THE PROCESS**

    A.   THE PROCESS
        1.   Unit Overview
        2.   Unit 100
        3.   Unit 200
        4.   Unit 300
        5.   Unit 400
        6.   Unit 500
        7.   Utilities
    B.   GEOGRAPHICAL LOCATION
    C.   CHEMISTRY OF THE PROCESS

III. **SAFETY, HEALTH AND ENVIRONMENTAL**

    A.   HAZARDOUS CHEMICALS
        1.   Properties and Hazards
        2.   Exposure Mechanisms
        3.   Prevention of Exposure
        4.   Response if Exposed to the Chemical
    B.   SLIPPING, FALLING, TRIPPING, CRUSHING
        1.   Types of Hazard and Their Potential
        2.   Prevention
        3.   Response
    C.   MATERIAL SAFETY DATA SHEETS (MSDS)
    D.   PERSONAL PROTECTIVE EQUIPMENT (PPE)
    E.   HOUSEKEEPING
        1.   General Rules
        2.   Inspection Teams

VII. **CONTACTS**
   A.   Fire and Ambulance
   B.   Environmental
   C.   Management
   D.   Vendors

VOLUME III — THE EMERGENCY OPERATIONS AND SHUTDOWN MANUAL

Most plants have three types of emergency manual.

1. The *Emergency Operations Manual*. This tells the operator how to run the plant when there is an emergency on his unit, or on another unit. This is an *operating* manual; it describes how to keep the plant running during an emergency.

2. The *Emergency Shutdown Manual*. This tells the operator how to bring the unit down as quickly as possible. This manual is to be used when there is a major catastrophe on the site.

3. The *Emergency Response Manual*. This is usually a site-wide document that is implemented only when the operators have completely lost control of their units, and control has passed to an external group, such as a fire brigade.

   Emergency procedures differ from normal operating procedures in the following ways.

- Some type of action *must* be taken; the option of doing nothing is not permitted.
- Response usually has to be very quick.
- The effectiveness of the response can truly be a life or death or issue. It is also likely to have major environmental economic impacts.
- Because there is an almost infinitely wide range of accidents that might occur, the emergency procedures have to be flexible. Yet, because time is of the essence, the procedures need to be rigid, short and precise. This is a difficult balance to strike.

Most emergencies do not occur in isolation. Usually, there is a multitude of events going on at once. For example, a fire is not likely to occur spontaneously. One of the reasons for its occurrence may have been loss of power to a critical compressor. The same loss of power may also have lead to a loss of fire fighting capability and a deterioration in the quality of communications, thus exacerbating the crisis.

Volume III is a set of operating instructions. Therefore it can be formatted in the same manner as Volume I. The only major difference occurs in the way that it is used. During an emergency an operator does not have time to read through a large amount of information. In fact, the emergency procedures should be short enough to memorize. Therefore the modules in Volume III should be very short indeed. (For example, there is no need for a Response/Discussion column.)

VOLUME IV — TROUBLESHOOTING MANUAL

Troubleshooting has been defined by Gans et al. as the "search for the hidden cause or causes that leads to inadequate performance." There are three key elements to "trouble" as distinct from an emergency.

1. It does not pose any safety problem, nor is there a significant chance of major economic loss
2. The response to trouble involves opinion, hunches and "feelings." This makes it very distinct from other types of operating procedure
3. Generally, time is not of the essence

Examples of "trouble" include:

1. Non-urgent product quality problems
2. Erratic machinery performance
3. Inexplicable environmental problems
4. Reduced yields of raw materials

If an operator does find that part of the process is out of control, he or she should consider the following questions.

1. Who do I call for help?
2. What do I do until help arrives?
3. How will other parts of the process (including other units) be affected by this problem?
4. What records should be kept for use in any subsequent incident investigation?

It has already been pointed out that there are many analogies between software and operating procedures. Troubleshooting presents another use for software concepts; in this case Expert Systems. These are programs that capture the experience of experts in a process in such a manner that it can be replayed by the operators who are on the spot. The hints and guidance of the Expert System may materially reduce the time it takes to solve the trouble because it often turns out that similar problems have occurred before.

VOLUME V — INITIAL START-UP MANUAL

Most people writing operating procedures are doing so for plants that are already running. Their assignment is to obtain the information that already exists on how to run the unit, and sort and organize it for the manual. In the case of brand-new plants, this is not the case. First, there is no existing operating experience on which to draw. It is possible that there are other, similar plants that can be used to provide "go-bys." Nevertheless, writing for procedures for new plants is inevitably more creative than it is for existing ones. The second problem is that the start-up process will have to start at an earlier point. Activities such as pre-commissioning and commissioning will have to be included. Normally, this is not necessary for existing plants.

However, there is one compensation for this increased difficulty: design conditions are well-defined and recorded. These represent the normal operating values, and can be used without further discussion (however, the operating limits still need to be developed.)

It is suggested that brand-new plants have a Volume V — The Initial Start-Up Manual. Information for it will typically come from three sources:

1. The licensor of the process
2. The engineering company that built the plant
3. The operating company's own standards and rules

It will typically contain three sections, none of which are found in Volume I. They are:

1. Mechanical completion. This is not usually the responsibility of the operations group. It often includes the hydrotest process.

2. Precommissioning. This usually includes very detailed punchlists that ensure that the plant is built as designed (this is at the heart of the Prestartup Safety Review element of the PSM standard.)

3. Commissioning. This covers equipment flushing, checking all rotating machinery and pressure testing (as distinct from hydrotesting.)

An important part of turning over a plant from construction to operations is the development of turnover packages. These divide the plant into discrete functional areas, such as the instrument air system or the boiler feed water system. These can be tested and punch listed when construction is finished while work is continuing in other areas. Turnover packages can also be used to organize the start-up manual. They are discussed on page 239.

THE WRITER'S GUIDE

Before a procedures-writing project starts, it is suggested that a Writer's Guide and a Project Manual be developed. The Writer's Guide provides instructions and assistance to those who are developing, writing and publishing the manual. Some of the issues to include in this Guide are described below.

PROTOTYPING

Although the top-down approach to developing a manual that was described in the first section is a good way of managing and controlling a procedures-writing project, it sometimes makes sense to develop one section of the manual in detail if a pilot project is needed.

For example, when a plant makes a commitment to completely re-design and re-write the operating procedures, they may choose to develop a prototype of a few modules for the following reasons:

1. It provides management with a better understanding as to how much time and effort will be needed for the complete project.

2. It gives everyone, particularly the operators, a chance to work with and to critique the format. It is much easier to comment on a worked example than to try and create something brand-new. If this step goes well, everyone is much more likely to buy into the overall project.

3. It highlights any problems that are likely to arise, such as missing information or difficulties with the information-gathering process.

PLAN TO THROW ONE AWAY

In his excellent book on the development of computer software — *The Mythical Man Month* — Brooks entitles one of his chapters: "Plan to throw one away; you will, anyhow." The idea behind this insight is that, no matter how well planned a software project may be, the changes that are made to a program during its design and development will be so significant as to make modifications of the original structure almost impossible. Instead, the software writers will do much better to discard what they have done and to start again, with all the changes in specification incorporated into the new design. Although the idea of throwing away the first product sounds draconian and wasteful, it is, in fact, far more productive than trying to fix the existing structure to incorporate major changes, and it causes fewer project management problems than might be expected.

This approach can be applied to the writing of operating procedures. Many procedures-writing projects reach a point at which so many changes are required, and there is so much re-design of the modules and their linkage to be done, that it is easier to start again. This does not mean that all the previous work is wasted; indeed, the opposite is the case. All of the information that has been gathered previously will be used, and will be connected into the system in a

useful and coherent manner. Usually the re-design focuses on the
linkage between the modules, rather than the modules themselves.

WRITING GUIDELINES

The following guidelines can be useful.

- Use the active voice
- Eliminate all redundant words
- List instructions singly, with a separate line for each
- Avoid use of the word "You"
- Avoid use of the word "should"

This last point is particularly important because it is a common
problem, and because it can lead to some serious misunderstandings.
For example, the phrase "The valve should be opened" could be
interpreted in the following ways:

1. Open the valve
2. You should already have opened the valve
3. Some other person or system should already have opened the valve
4. You might choose to open the valve, depending on other
   conditions
5. We suggest that you open the valve

THE USE OF COLOR

OSHA has a standard for signs and color: 29 CFR 1910.144.
Basically, the standard is as follows:

- *Red* is the basic color of identification of fire protection equipment
  and apparatus. It is also used for safety cans or other portable
  containers of flammable liquids. (Identification should be made
  with a yellow band.) Red lights are to be used at barricades and
  temporary barricades. It is also to be used for stop bars on
  hazardous machines and stop buttons.

- *Yellow* is the color for designating caution and for marking
  physical hazards.

- *White* is used for notices and informational signs. It should have black letters on it.

SAMPLE WRITER'S GUIDE

Table 6-3 provides a representative Table of Contents for the Writer's Guide.

Table 6-3

Representative Table of Contents for The Writer's Guide

I.   INTRODUCTION
II.   MANUAL ADMINISTRATION
    A.   MANUAL ORGANIZATION
        1.   Procedures
        2.   Information
        3.   Emergencies and Trouble-Shooting
    B.   VOLUME I — OPERATING PROCEDURES
        1.   Modular Structure
        2.   Design of a Module
        3.   Linkage of Modules
    C.   VOLUME II — OPERATING INFORMATION
    D.   VOLUME III — EMERGENCY OPERATING AND SHUTDOWN PROCEDURES
    E.   VOLUME IV — TROUBLESHOOTING GUIDE
    F.   VOLUME V — INITIAL START-UP
    G.   DISTRIBUTION
    H.   MANUAL REVISION
    I.   ANNUAL CERTIFICATION

III.   COMPUTER HARDWARE/SOFTWARE
    A.   COMPUTER HARDWARE
        1.   Computers
        2.   Printers
        3.   Plotters
        4.   Scanners/Digital Imaging
    B.   COMPUTER SOFTWARE
        1.   Word Processing
        2.   Graphics

THE PROJECT MANUAL

The Project Manual provides guidance as to how a procedures-writing project should be organized and controlled. This means that management must treat the writing of procedures as being a project equivalent to say installing a large item of equipment. Therefore, there needs to be a budget, a schedule and a project manager.

It is suggested that operating procedures should be written by teams, rather than by a single individual who is responsible for all aspects of the project. The team approach will ensure that specialist skills are utilized most effectively; also interaction between the team members should improve the quality of the product, and it will maximize employee participation. Therefore, the first step in managing a procedures-writing project is to decide on the function and composition of the teams. The approach described in this chapter is to use two teams. The first of these, *The Resource Team*, is responsible for developing the technical information that is to be included in the manual. The second team, *The Publishing Team*, is responsible for editing and publishing the final product. If a company decides to use a contractor to help with the project, the contractor will most likely fill the role of the Publishing Team and/or the project management. Both of these teams report to a single project manager, who is in overall control of the project. The organization chart for this management system is shown in Figure 6-8.

Figure 6-8

Representative Organization Chart

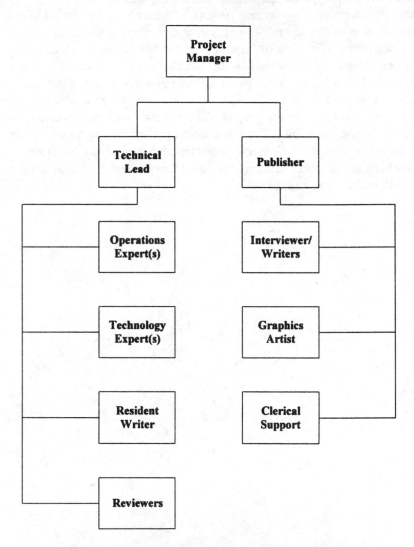

The project manager is in charge of the overall project. He is responsible for all facets of the work, including technical quality, financial control and scheduling. He does not necessarily have to be an expert on procedures or on the unit being described.

In practice, the overall chart will probably be somewhat smaller, with one person filling two or more of the roles. But, they should always understand the distinct nature of these roles.

STANDARD OPERATING PROCEDURES

Earlier in this chapter, the concept of Standard Operating Procedures (SOP's) was introduced. The basic idea is that there many operations are very similar to one another, hence it makes sense to develop standard or generic procedures, that can then be used as the basis for specific procedures. These SOP's can be very carefully checked and reviewed before being published. The advantages of doing this are:

- There will be a substantial saving of time due to the fact that many of the procedures will have been largely pre-written.
- The procedures will be of high quality. In particular, there will be less chance of some steps being overlooked.
- Problems associated with writer's block will be minimized.

Some representative titles for SOP's are provided in Table 6-4. This list is not intended to be comprehensive, but it does provide some guidance as to how a Table of Contents could be developed. The items in Table 6-4 are all equipment oriented. Another way of developing SOP's is for types of technology, such as refineries or food processing plants.

Table 6-4

Representative Standard Operating Procedures

100 Normal Operations

Tab 101 — Routine Rounds
    101.01 — Outside Readings
    101.02 — Inside Readings

300 Start-Up

400 Shutdown

500 Emergency Operations

Tab 501 — Emergency Shutdown

Tab 502 — Emergency Response
    501.02 — Fires and Explosions
    501.03 — Release Of Toxic Gas

Tab 510 — Response Following An Emergency
    502.01 — Clean-Up Following An Emergency
    502.02 — Start-Up Following An Emergency Shutdown

600 Troubleshooting

Tab 601 — Compressors
    601.01 — Reciprocating Compressors (1 of 2)
    601.02 — Reciprocating Compressors (2 of 2)
    601.05 — Centrifugal Compressors

Tab 602 — Pumps
    602.01 — Centrifugal Pumps (1 of 3)
    602.02 — Centrifugal Pumps (2 of 3)
    602.03 — Centrifugal Pumps (3 of 3)

700 Operations / Maintenance

Tab 701 — Valves and Piping
    701.01 — Putting A Relief Valve In Service
    701.02 — Taking A Relief Valve Out Of Service

Tab 702 — Vessels

Tab 703 — Pumps

Tab 704 — Compressors/Fans

Tab 705 — Instrumentation/DCS

800 Miscellaneous

Tab 801 — Vehicle Movement
    801.01 — Verification Of Truck Documents
    801.04 — Loading A Truck
    801.05 — Unloading A Truck
    810.01 — Changing Filters
    820.01 — Pigging A Line

CERTIFICATION

OSHA requires that the procedures be certified every 12 months in their knowledge of operating procedures. This does not mean that the procedures have to be revised — unless there is a change in the operation — but it does mean that each procedure has to be checked as to its current validity. Table 6-5 provides an example of a Certification Form.

Table 6-5

Sample Certification Form

Time Period Covered
Date Of Last Certification _____
Date Of This Certification _____

Process Covered _____
_____
_____

Procedures Covered _____
_____
_____
_____
_____

I certify to the best of my knowledge as of the date below that the written operating procedures for processes covered by OSHA Standard 29 CFR 1910.119 f(3) provide clear and current instructions for safely conducting activities involved in the processes.

Name _____ (Print or type)

Title _____ (Print or type)

Date _____ (Print or type)

Signature _____

MAINTENANCE PROCEDURES

OSHA does not have a section in the standard explicitly to do with maintenance procedures. Nevertheless, maintenance procedures are certainly an important part of a complete process safety system, and they are alluded to in the parts of the regulation to do with PHA's, Mechanical Integrity and Management of Change.

Many of the ideas and concepts presented here with regard to operating procedures can be used for maintenance procedures. However, there are some differences. Often operating procedures involve a series of sequential tasks, and the order in which those tasks is carried out is critical. This is not so true with regard to maintenance procedures, which tend to be more concerned with individual, one-off activities, such as repairing a piece of equipment.

In terms of their appearance, maintenance procedures can make much more use of photographs and sketches than can operating manuals. In particular, exploded views of the items being worked on can be very useful.

ELECTRONIC OPERATING PROCEDURES

ON-LINE MANUALS

 Traditionally, Operating Manuals have been published on paper, with loose-leaf notes inserted in a 3-ring binder. The ease with which pages can be added and removed is important; it allows for updates to be made easily and quickly. However, there are drawbacks to this way of publishing. Such drawbacks include:

1. Difficulties of keeping all copies up to date. If a number of copies of the manual have been issued, it can be difficult to find them when the time comes to update them.

2. Difficulties with physical integrity. Paper manuals can be damaged easily. Sometimes pages will be torn out or damaged accidentally. On other occasions, someone may remove a page for use in the field, and then forget to return it.

3. Limited use of color. Color printing is difficult and expensive. Consequently, most manuals are printed in black and white. Even if color is used, they are frequently copied on normal photocopy machines, which remove all color.

4. Static Imagery. Paper manuals do not allow movement to be shown; all the pictures and graphics are static only.

One way around these problems is to use of local area networks, in which the manual resides on a central server. The users all have access to terminals. If they need instructions, they can read them on the screen or they can print them out. If they print them, each page will be stamped with a termination date and time (giving the procedure a "life" of say 24 hours.) After that time, the printed procedure is no longer valid, and the operator will have to print it out again if he/she wants to perform that particular task again. The advantages of this type of system are:

1. There are no distributed copies to update. The person responsible for issuing new procedures merely has to update the copy on the server. There is no need to circulate pieces of paper.

2. If an operator prints out an instruction set, and then loses it or damages it, all that he/she has to do is print out another copy.

3. Color can be used on the screen. Also, there is no limit to the number of graphics and photographs that can be added.

4. It is feasible to add moving pictures, movie clips and sound.

None of the above ideas are new. However, recent commercial software from well-known companies has made this technology widely available, easy to use and affordable. The steady growth of network systems in most companies also contributes toward the feasibility of this way of distributing operating procedures.

OSHA has expressed concern about the reliability of an electronic manual in the event of a power failure or computer system crash. Therefore, it is probably necessary to keep one up-to-date copy of the manual in the control room. However, this concern may disappear once electronic manuals become more common place.

HYPERLINKS

Hyperlinks allow documents to call on one another. Once more, this technology has been in existence for some years. However, the rapid growth of the World Wide Web in the late 1990's has made it part of normal business. In the case of procedures, hyperlinks provide an ideal way for an operator to pull up more information when he/she needs it; this technology helps solve the problem of how much detail to put in the procedures. The top-level procedure can be very brief, essentially in checklist form. Then, if the operator wants help, all that he/she has to do is to jump to other documents which can provide the necessary information.

For example, the simple instruction "Start Pump, P-100" could be divided into two sections: "Start Pump" and "P-100". Clicking on the phrase "Start Pump" would take the operator to another on-line document entitled "How To Start A Pump." This document would provide general information on the procedures for starting pumps. Clicking on the word "P-100" would pull up a document "Details About Pump P-100." This document would provide technical information about that pump (size, capacity, materials of construction, and so on.) In each case, once the operator had read the supplemental document, he/she could flip back to the original document "Start Pump, P-100."

Once a system such as this is in place, the "Response/Discussion" column can be removed from the Operating Procedures module, thus reducing the size of the manual. This is very much in the spirit of minimalist writing.

These links would not be restricted to documents on the same computer or server as the user. They could connect to other networks, including the Internet.

CONCLUSIONS

Operating procedures are a critical part of any process safety program because they define the relationship between the operators and the equipment. Hence almost all other elements have an effect on the procedures.

The development of good procedures requires a well-thought out organization and a top-down approach to their development. Otherwise, the effort required can make the task of writing procedures seem monumental.

Chapter 7

TRAINING

INTRODUCTION

 All aspects of process safety come down to people working with systems; if the people are not properly trained in what to do, then the systems for which they are responsible will not be safe. Training can be divided into one of three types:

- Introductory or generic
- Site
- Task specific

Introductory training is designed for new employees, people who may have no knowledge or background of any kind in how process plants work. It teaches people the basic operations (such as how to open a valve or turn on a pump.) It also covers standard safety practices. This type of training is not specific to any company, facility or technology. It provides the operators with the rudimentary skills and knowledge that they need to work on a process plant.

Site training instructs the operators in how to work on a particular site. Employees are instructed in company practices, such as the management of work permits, and on general policies, such as response to emergencies.

Task-specific training instructs the operators on how to perform specific tasks to do with their specific assignments. It can be closely linked with the modular approach to operating procedures described in the previous chapter.

It is useful to draw a distinction between training and education. Training shows a person how to do a task, education explains the principles that lie behind the task. For example, if an operator is learning how to start a pump, then he is being trained. However, if the

212 Process Safety Management

principles of pump operation are being explained, he is being educated. Both training and education are needed. A person has to be trained in the safe execution of his or her day-to-day tasks. However, education is needed if they are to assist with trouble-shooting and other non-standard activities.

THE OSHA REGULATION

(1) Initial training.

(i) Each employee presently involved in operating a process, and each employee before being involved in operating a newly assigned process, shall be trained in an overview of the process and in the operating procedures as specified in paragraph (f) of this section. The training shall include emphasis on the specific safety and health hazards, emergency operations including shutdown, and safe work practices applicable to the employee's job tasks.

(ii) In lieu of initial training for those employees already involved in operating a process on May 26, 1992, an employer may certify in writing that the employee has the required knowledge, skills, and abilities to safely carry out the duties and responsibilities as specified in the operating procedures.

(2) Refresher training. Refresher training shall be provided at least every three years, and more often if necessary, to each employee involved in operating a process to assure that the employee understands and adheres to the current operating procedures of the process. The employer, in consultation with the employees involved in operating the process, shall determine the appropriate frequency of refresher training.

(3) Training documentation. The employer shall ascertain that each employee involved in operating a process has received and understood the training required by this paragraph. The employer shall prepare a record which contains the identity of the employee, the date of training, and the means used to verify that the employee understood the training.

OSHA GUIDANCE

All employees, including maintenance and contractor employees, involved with highly hazardous chemicals need to fully understand the safety and health hazards of the chemicals and processes they work with for the protection of themselves, their fellow employees and the citizens of nearby communities. Training conducted in compliance with 1910.1200, the Hazard Communication standard, will help employees to be more knowledgeable about the chemicals they work with as well as familiarize them with reading and understanding MSDS. However, additional training in subjects such as operating procedures and safety work practices, emergency evacuation and response, safety procedures, routine and nonroutine work authorization activities, and other areas pertinent to process safety and health will need to be covered by an employer's training program.

In establishing their training programs, employers must clearly define the employees to be trained and what subjects are to be covered in their training. Employers in setting up their training program will need to clearly establish the goals and objectives they wish to achieve with the training that they provide to their employees. The learning goals or objectives should be written in clear measurable terms before the training begins. These goals and objectives need to be tailored to each of the specific training modules or segments. Employers should describe the important actions and conditions under which the employee will demonstrate competence or knowledge as well as what is acceptable performance.

Hands-on-training where employees are able to use their senses beyond listening, will enhance learning. For example, operating personnel, who will work in a control room or at control panels, would benefit by being trained at a simulated control panel or panels. Upset conditions of various types could be displayed on the simulator, and then the employee could go through the proper operating procedures to bring the simulator panel back to the normal operating parameters. A training environment could be created to help the trainee feel the full reality of the situation but, of course, under controlled conditions. This realistic type of training can be very effective in teaching employees correct procedures while allowing them to also see the consequences of what might happen if they do not follow established operating procedures. Other training techniques using videos or on-the-job training can also be very effective for teaching other job tasks, duties, or other important information. An effective training program will allow the employee to fully participate in the training process and to practice their skill or knowledge.

Employers need to periodically evaluate their training programs to see if the necessary skills, knowledge, and routines are being properly understood and implemented by their trained employees. The means or methods for evaluating the training should be developed along with the training program goals and objectives. Training program evaluation will help employers to determine the amount of training their employees understood, and whether the desired results were obtained. If, after the evaluation, it appears that the trained employees are not at the level of knowledge and skill that was expected, the employer will need to revise the training program, provide retraining, or provide more frequent refresher training sessions until the deficiency is resolved. Those who conducted the training and those who received the training should also be consulted as to how best to improve the training process. If there is a language barrier, the language known to the trainees should be used to reinforce the training messages and information.

Careful consideration must be given to assure that employees including maintenance and contract employees receive current and updated training. For example, if changes are made to a process, impacted employees must be trained in the changes and understand the effects of the changes on their job tasks (e.g., any new operating procedures pertinent to their tasks). Additionally, as already discussed the evaluation of the employee's absorption of training will certainly influence the need for training.

ANALYSIS OF THE REGULATION

The OSHA regulation with regard to Training is quite straightforward. Basically, operators must be trained in the operating procedures for the unit in which they are working, and their training must be updated as needed, but always within three years. The "grandfather" clause for those operators who were working on the facility before the standard was issued is becoming less relevant as new operators come on board, and as experienced operators go through the three year refresher programs. This topic of "grandfathering" can be sensitive. Management may suspect that some of the experienced operators do not fully understand the tasks for which they are responsible, and which they are currently performing. Therefor they can use the training program to (a) test everyone and determine if their suspicions are correct, and (b) bring everyone up to the same standard. However, the operators may resent this; they may feel that they are being "put down" by being tested on actions that they feel they know very well indeed.

The OSHA Guidance refers to the 1910.1200 (HAZWOPER) standard as a source of training, particularly with regard to hazardous chemicals.

In terms of meeting the OSHA standard, documentation of training is often a major problem. Companies may sponsor training programs of all types, but, unless the training is properly documented, it will not be recognized as satisfactory by an auditor.

There is a very close link between Operating Procedures and Training. However, they are not the same: procedures are written for operators who have already been trained. Before the operator can use the procedures, he or she needs to be trained in the operation.

Training can be integrated with the modular concept of operating procedures (*see* page 175.) Along with the Action and Response columns, there can be two training columns: one of them called "Why?" and the other "How?" The "How?" column provides more detail as to how to do a task. The Response/Discussion column alluded to in the previous chapter can fill this role. The "Why?" column provides an explanation, and so is education rather than training.

One way of linking procedures with training is to provide a training video shows an experienced operator actually working through the steps in the appropriate module. While describing what he was doing for each of the steps, the words and actions would be recorded on a video tape. The tape would then be kept in the control room so that the operators could review it at any time. This could also be used to describe incidents that have occurred in the past relating to this step and he could also use it to show visually any difficulties or special features of the operation. It would provide a fine foundation for a trouble-shooting program.

IMPLEMENTING A TRAINING PROGRAM

BASIC/GENERIC TRAINING

Orientation of new employees starts on the first day of work. Items covered include: plant organization, departmental functions, chemicals used, safety, environmental, maintenance, work rules and general responsibilities. The orientation will be provided by representatives from the appropriate departments. The safety training will emphasize the need to report injuries, the use of Personal Protective Equipment, and response to emergencies.

Regulatory training covers environmental policies and procedures, OSHA regulations to do with hot work, lockout/tagout, confined space entry and hearing protection, and HAZWOPER and HAZCOM requirements.

The plant basic training will explain how equipment such as pumps, valves, electrical motors, instruments and heat exchangers work. It will also cover basic technologies, such as chemistry and distillation (where needed.) This training provides an overview of how the plant works, what it makes, and what its safety, environmental and economic objectives are.

Basic training is what an operator has to have before he or she is permitted to carry out any task, no matter how simple. The nature of the basic training will naturally vary from plant to plant. However, a program such as the one described below is representative.

- *Safety Policies.* This will principally be occupational or "hard hat" safety (*see* page 12), rather than an understanding of process safety management issues.

- *Emergency Response and Emergency Operations.* The new operator will be made familiar with the plant emergency systems, and what his or her role will be should an emergency occur. Training as to how to run particular units during an emergency will come later.

- *Company Structure.* This part of the training will provide an overview of the different departments at the site, and what their responsibilities are. It will also explain the lines of authority, both in normal operation and at times of emergency.

- *Permits To Work.* The basic training will include a thorough explanation of the Permit to Work system, including lockout/tagout, hot work procedures and vessel entry procedures. The training should focus on the interface between operations and maintenance. It is essential that the operator trainee knows where the lines of responsibility lie.

- *Special Safety Equipment.* The operator will be introduced to special safety equipment, such as self-contained breathing apparatus (SCBA) and chemically-resistant clothing, and be trained in its use.

- *Coordination with Other Departments.* The role of other departments, particularly maintenance, instrumentation and safety, will be explained. The way in which operations interacts with these departments, and who is responsible for what, is described.

PLANT TRAINING

Once an operator has received basic orientation, the next step is to provide him or her with plant training. The goal is to familiarize him with the processes, how they work, what they make, and how they inter-act with one another. This part of the training program will be even more plant and company-specific than the orientation.

This will be conducted by an experienced operations person such as a supervisor or a senior operator. The trainee will be instructed on details as to how a particular unit works. Daily training reports should be prepared. These describe what the trainee has covered, and lists any test results. At the conclusion of this phase, the trainee will be tested by supervision and management. If he or she passes, they are qualified to work on that unit.

The elements outlined below are representative of Plant and On-The-Job Training.

- *Plant or Refinery Overview.* This provides the trainee with an understanding of the overall process.

- *Major Equipment.* This section of the training will provide an overview of how to equipment items work, and how to operate them. The equipment covered will typically include pumps, heat exchangers, distillation columns, compressors and valves. Generic procedures will include items such as starting pumps, opening and closing valves and catching samples.

- *Emergency Response.* The training must cover the emergency response systems that are used within the facility.

- *Instrumentation and Safety Systems.* The trainee will be provided with an explanation of the plant's instrumentation, computer control systems and emergency shut down systems. The principles of measurement and control, including proportional, differential and integral controllers, will be explained.

- *Regulations.* The trainee will be introduced to the more important safety and environmental regulations, and told what he or she is expected to do to conform to those regulations. These are likely to include the HAZMAT and HAZWOPER programs.

TESTING

Once an operator has completed a plant training program, he or she has to be tested to make sure that they understood what they were taught. Testing is not an option; only if people pass the appropriate tests can management be sure that their employees know enough to operate the plant safely.

Testing can be formal, for example by using multiple-choice questions from a book. It can also be informal. Some supervisors include in their tests a simple walk around the unit with the trainee operator. During this walk around, the operator is asked a series of questions in which he or she has to show what has to be done in various situations. In one case, a company had a number of very simple tests, many of which were known to the trainees ahead of time. For example, the trainees would be asked to draw from memory on a blank piece of paper a plot plan of their area, with the location of all safety and fire-fighting equipment identified.

CERTIFICATION

Once the employee has passed the tests for an area and for a job class, the facility may choose to certify him or her. The exact meaning of the word "certification" in the context of process safety is open to interpretation, and will vary from site to site. It can also have legal implications, for example in the negotiation of union contracts.

CONCLUSIONS

Training is an integral part of process safety; if people do not know what to do, how to do it, and why they are doing it, accidents will happen. The systems for conducting training are generally well established. What is needed is to link training with the other elements of PSM, particularly Employee Participation and Operating Procedures.

Chapter 8

CONTRACTORS

INTRODUCTION

Contract workers are used for many of the activities covered by process safety, particularly in the areas of maintenance and construction. The intent of this section of the standard is to ensure that contract workers observe the same safety rules and standards as permanent employees.

There are at least four issues to be considered when developing a contractor safety program.

1. Traditionally, a contract worker is someone who is brought on site for a short period of time, completes a pre-defined task, and then leaves. However, in many plants, contract workers now work full-time at the site. Consequently some of these workers will have years of service at a particular site, and may indeed have a good deal more knowledge regarding that operation than many of the full-time employees. Hence, it is important not to think of the contract workers as outsiders who need to be trained in the facility's process safety program, but as full members of the team, who may well contribute more to the program than they receive because of their detailed knowledge of actual plant practices.

2. Contractors are often involved in some of the most hazardous activities on the plant because they frequently work in maintenance and construction activities. Hence they routinely carry out tasks such as entering vessels, repairing high voltage electrical equipment and operating cranes. Yet, because they may not be present at the site full time, contract workers may not be familiar with the safety procedures to be followed.

3. Contract workers often have two sets of managers: the facility management and the contract employer. This means that there is the potential for divided responsibility, particularly in those cases

where the contract workers are on site for a long period of time and report directly to the host company. In these cases, the contracting company serves almost like an employment agency.

4. Those contract workers who are present on the site for only a short time, will not be familiar with the facility's overall safety culture, nor with its internal organization. This can lead to misunderstandings and faulty communication.

The OSHA standard to do with contractors is concerned primarily with these management issues. The technical aspects of contractor process safety are basically the same as for full-time employees. Problems are most likely to arise due to misunderstandings as to who is responsible for the safety and actions of the individual contract workers.

THE OSHA REGULATION

(1) Application. This paragraph applies to contractors performing maintenance or repair, turnaround, major renovation, or specialty work on or adjacent to a covered process. It does not apply to contractors providing incidental services which do not influence process safety, such as janitorial work, food and drink services, laundry, delivery or other supply services.

(2) Employer responsibilities.

(i) The employer, when selecting a contractor, shall obtain and evaluate information regarding the contract employer's safety performance and programs.

(ii) The employer shall inform contract employers of the known potential fire, explosion, or toxic release hazards related to the contractor's work and the process.

(iii) The employer shall explain to contract employers the applicable provisions of the emergency action plan required by paragraph (n) of this section.

(iv) The employer shall develop and implement safe work practices consistent with paragraph (f)(4) of this section, to control the entrance, presence and exit of contract employers and contract employees in covered process areas.

(v) The employer shall periodically evaluate the performance of contract employers in fulfilling their obligations as specified in paragraph (h)(3) of this section.

(vi) The employer shall maintain a contract employee injury and illness log related to the contractor's work in process areas.

(3) Contract employer responsibilities.

(i) The contract employer shall assure that each contract employee is trained in the work practices necessary to safely perform his/her job.

(ii) The contract employer shall assure that each contract employee is instructed in the known potential fire, explosion, or toxic release hazards related to his/her job and the process, and the applicable provisions of the emergency action plan.

(iii) The contract employer shall document that each contract employee has received and understood the training required by this paragraph. The contract employer shall prepare a record which contains the identify of the contract employee, the date of training, and the means used to verify that the employee understood the training.

(iv) The contract employer shall assure that each contract employee follows the safety rules of the facility including the safe work practices required by paragraph (f)(4) of this section.

(v) The contract employer shall advise the employee of any unique hazards presented by the contract employer's work, or of any hazards found by the contract employer's work.

OSHA GUIDANCE

Employers who use contractors to perform work in and around processes that involve highly hazardous chemicals, will need to establish a screening process so that they hire and use contractors who accomplish the desired job tasks without compromising the safety and health of employees at a facility. For contractors, whose safety performance on the job is not known to the hiring employer, the employer will need to obtain information on injury and illness rates and experience and should obtain contractor references.

Additionally, the employer must assure that the contractor has the appropriate job skills, knowledge and certifications (such as for pressure vessel welders). Contractor work methods and experiences should be evaluated. For example, does the contractor conducting demolition work swing loads over operating processes or does the contractor avoid such hazards?

Maintaining a site injury and illness log for contractors is another method employers must use to track and maintain current knowledge of work activities involving contract employees working on or adjacent to covered processes. Injury and illness logs of both the employer's employees and contract employees allow an employer to have full knowledge of process injury and illness experience. This log will also contain information which will be of use to those auditing process safety management compliance and those involved in incident investigations.

Contract employees must perform their work safely. Considering that contractors often perform very specialized and potentially hazardous tasks such as confined space entry activities and nonroutine repair activities it is quite important that their activities be controlled while they are working on or near a covered process. A permit system or work authorization system for these activities would also be helpful to all affected employers. The use of a work authorization system keeps an employer informed of contract employee activities, and as a benefit the employer will have better coordination and more management control over the work being performed in the process area.

A well run and well maintained process where employee safety is fully recognized will benefit all of those who work in the facility whether they be contract employees or employees of the owner.

ANALYSIS OF THE REGULATION

There are four basic elements of the standard to do with contract workers:

1. The contract worker must be trained in all issues to do with hazardous chemicals as they relate to the work he is doing.

2. The facility management must satisfy themselves that the contracting company operates to sufficiently high safety standards.

3. The facility management must make sure that all necessary information is made available to the contracting company and to the contract workers.

4. The contracting company must ensure that the individual contract workers are properly trained, and that records are kept.

(1) Application

This paragraph provides guidance as to what constitutes a contract worker. Sub-contractors and their employees are included.

Although OSHA separates contract workers who are providing "incidental" services from those who are working with hazardous chemicals, it is important to ensure that their work is indeed incidental to the process, and that they do not become inadvertently involved with the process in some manner. For example, if a contract worker is supplying snacks and soft drinks for the lunch room inside a plant, he or she will probably need to know about the use of basic safety clothing, and what to do in the event of an emergency.

(2) Responsibilities

This paragraph highlights what the employer is expected to do when hiring and training contractors. Note that all of these requirements explicitly require the employers to be involved in the management of contractors. Companies can not distance themselves from safety responsibilities by handing off the work to an outside contractor, and then "leaving them to it."

The standard makes it clear that a contract company's safety record is to be reviewed before they are hired, and that safety record should be evaluated as part of the overall contractor selection process. Generally the contractor's OSHA 200 logs will be used for this evaluation.

Carmel provides guidance concerning the selection of a contractor. Some of the suggested areas to look for include the following:

- Check the contractor's safety logs, and compare them with the OSHA logs
- Check for willful violations
- Check for repeat violations
- Check for violations in areas of expertise
- Check the training program and training records
- Check Incident Investigation reports

If a contracting company consistently falls short in one of these areas it should be disqualified from further work at that site.

This paragraph also requires that contract employees be fully informed as to hazards and potential emergency situations. This will usually include an overview of the entire facility, the process on which the worker is working, and the relevant fire, explosion and toxic hazards. Contractors workers should know what to do when an emergency has been announced. In particular, they must know what to do with power tools that are in use, and where the evacuation routes are. Key elements of this training will typically include the following:

1. Emergency recognition and prevention
2. Safe distances
3. Places of refuge
4. Evacuation routes
5. Site security and control
6. Decontamination procedures
7. Emergency medical treatment and first aid
8. Emergency alerting and response procedures
9. Personal Protective Equipment
10. Emergency equipment

The employer is responsible for the development and implementation of safe work practices. In many cases, the contract company will also have its own standard practices. Any differences between the two companies should be resolved before work starts.

Employers are required to periodically evaluate and audit contract employers. If any problems or violations of company standards are observed, or if the contractor's safety record is deteriorating, the operating company must bring them to the attention of the contractors, and ensure that these problems are addressed.

(3) Contract Employer

The contract employer is responsible for the detailed implementation of the programs provided by the facility. With regard to training, the company will usually have a two-part program. The first part involves the general safety policies and procedures, to be followed at all times. The second part of the training will be site-specific; the information for this will be provided by the facility. The content of this course will include issues discussed above.

DEFINITION OF A CONTRACTOR

From OSHA's point of view, the key factor in determining whether a person who is working at a facility is a contractor or not is who does the day-to-day supervision. If this supervision is mostly performed by the facility personnel, then the "contract" workers will be treated as if they are full-time employees. If, on the other hand, the contract workers are supervised by their own company, and if they have little interaction with the facility's process, then they would not be regarded as full-time employees. For example, if there is a construction crew on site erecting a new administration building, the workers who pour the concrete and erect the steel work will not have any contact with the people who work on the processes. Therefore, they are true "contract workers." At the other extreme is a maintenance worker who has worked on the facility for many years and who reports directly to the facility's supervision. Even though this worker may be contract in an administrative sense, OSHA would regard him/her as equivalent to an employee in terms of process safety. The rationale behind this way of defining a contractor is that "OSHA believes that whoever is telling the work what to do is probably in the best position to protect the worker." (Threet.)

RECORDKEEPING

OSHA 200 LOGS

As with other elements of PSM, it is important that good records be maintained and used. When an employer in the United States is evaluating a contractor, the basic safety data will usually come from OSHA 200 logs (although other types of safety logs can be used.) An explanation of the OSHA 200 system is provided in their standard 29 CFR 1904.2. Details are to be found in the Supplementary Record of Occupational Injuries and Illnesses, OSHA 101, a summary of which is provided below.

(a) Each employer shall, except as provided in paragraph (b) of this section, (1) maintain in each establishment a log and summary of all recordable occupational injuries and illnesses for that establishment; and (2) enter each recordable injury and illness on the log and summary as early as practicable but no later than 6 working days after receiving information that a recordable injury or illness has occurred. For this purpose form OSHA No. 200 or an equivalent which is as readable and comprehensible to a person not familiar with it shall be used. The log and summary shall be completed in the detail provided in the form and instructions on form OSHA No. 200.

(b) Any employer may maintain the log of occupational injuries and illnesses at a place other than the establishment or by means of data-processing equipment, or both, under the following circumstances:

(1) There is available at the place where the log is maintained sufficient information to complete the log to a date within 6 working days after receiving information that a recordable case has occurred, as required by paragraph (a) of this section.

(2) At each of the employer's establishments, there is available a copy of the log which reflects separately the injury and illness experience of that establishment complete and current to a date within 45 calendar days.

Further information regarding OSHA 200 logs is given in the Recordkeeping Guidelines (OSHA 1986.) Some of the key statements from these Guidelines are quoted below.

A. The Log of Occupational Injuries and Illnesses, OSHA No. 200

The log is used for recording and classifying recordable occupational injuries and illnesses, and for noting the extent and outcome of each case. The log shows when the occupational injury or illness occurred, to whom, what the injured or ill person's regular job was at the time of the injury or illness exposure, the department in which the person was employed, the kind of injury or illness, how much time was lost, and whether the case resulted in a fatality, etc.

The log consists of three parts: A descriptive section which identifies the employee and briefly describes the injury or illness; a section covering the extent of the injuries recorded; and a section on the type and extent of illnesses. [A standard OSHA 200 form is used by most employers.]

Employer Decisionmaking

Employers decide which cases are to be entered on the OSHA records. This decision must be made in good faith, according to the requirements of the act.

Analysis of Recordability

Recording an injury or illness under the OSHA system does not necessarily imply that management was at fault, that the worker was at fault, that a violation of an OSHA standard has occurred, or that the injury or illness is compensable under workers' compensation or other systems.

RECORDABLE CASES: You (the employer) are required to record information about every occupational *death*; every nonfatal occupational *illness*; and those nonfatal occupational *injuries* which involve one or more of the following: loss of consciousness, restriction of work or motion, transfer to another job, or medical treatment (other than first aid.)

The OSHA 200 logs must be filed with OSHA, and must be posted by February 1 of the following year in an area frequented by employees until March 1.

OSHA RECORDABLES

The OSHA Recordable Rate is calculated for the previous three years. It is defined as:

$$\frac{\text{Number of Recordable Cases} \times 200,000}{\text{Total Hours Worked}}$$

The OSHA Lost Workday Incident Rate is similar:

$$\frac{\text{Number of Lost Workday Cases} \times 200,000}{\text{Total Hours Worked}}$$

A lost workday — equivalent to a lost time injury — is one where an individual misses more than one day of work due to an injury sustained while at work. In order to avoid a lost-time injury, the worker must report to work for his or her next scheduled shift following the occurrence of the injury.

EXPERIENCE MODIFICATION RATE

Knowledge as to a contractor's performance can be provided by their insurance company. They calculate an Experience Modification Rate (EMR) for the previous three years. It is the ratio of the "Actual Workers Compensation Losses" to the contractor's "Expected Workers Compensation Losses."

Expected losses are determined from published tables which state the "average losses per $100 of payroll" for each of the different manual rating classifications and the actual payroll for that classification. Actual losses of the money paid, plus the money estimated to be paid, by the insurance company for injuries and include medical treatment, wage compensation, and other treatment compensation provided by the workers compensation system. A 1.0 EMR is assigned to the "average" contractor in their particular type of business.

CONCLUSIONS

The management of contractors is critically important to process safety. Contract workers are often involved in some of the most hazardous tasks on the unit, and they do not know the company's culture and standards. The OSHA regulation makes it clear that both the management of the facility and the contracting company are responsible for making sure that each contract worker is fully trained and has a well-defined line of reporting authority.

Chapter 9

PRESTARTUP SAFETY REVIEW

INTRODUCTION

SLOW

The purpose of a Prestartup Safety Review (PSSR) is to ensure that initial start-ups, or start-ups following major turnarounds, are carried out safely. In particular, a PSSR aims to confirm that what was built and installed conforms to the original design and standard, and that no hazardous situation has been created during construction and commissioning. (If the review is being carried out on a plant that is being re-started following a turnaround, it is referred to as a Restart Safety Review.) It provides a breathing space for everyone to make sure that the plant that they are about to start up is safe, and that no safety corners have been cut.

PSSR's are important because projects frequently fall behind schedule and/or run over budget. This creates pressure on the project team to eliminate or postpone the installation of any items that are not absolutely necessary for the start-up. If not controlled properly, this can lead to corner-cutting — either intentional or inadvertent — which may in turn jeopardize the safety of the plant. By carrying out a PSSR, the operations department has the authority to refuse to accept "care, custody and control" of a plant that they judge to be unsafe.

PSSR covers not only equipment, but also human factors issues such as operating procedures and training. This is particularly important with regard to Restart Reviews; it is vital that the operating procedures be updated to reflect the changes that have been made, and that the operators are then trained in the new procedures before they start work on the modified facility.

PSSR's frequently identify documentation problems with respect to process safety. During the construction and commissioning of a plant, there is usually considerable pressure just to get the plant completed and up and running. Hence, the record-keeping part of the project may slip. If, in the judgment of the PSSR team, some of the missing documents are important to safety, they can must ensure that those documents are completed and issued before the plant is started.

THE OSHA REGULATION

The employer shall perform a pre-startup safety review for new facilities and for modified facilities when the modification is significant enough to require a change in the process safety information.

The pre-startup safety review shall confirm that prior to the introduction of highly hazardous chemicals to a process:

(i) Construction and equipment is in accordance with design specifications;

(ii) Safety, operating, maintenance, and emergency procedures are in place and are adequate;

(iii) For new facilities, a process hazard analysis has been performed and recommendations have been resolved or implemented before startup; and modified facilities meet the requirements contained in management of change, paragraph (l).

OSHA GUIDANCE

For new processes, the employer will find a PHA helpful in improving the design and construction of the process from a reliability and quality point of view. The safe operation of the new process will be enhanced by making use of the PHA recommendations before final installations are completed. P&ID's are to be completed along with having the operating procedures in place and the operating staff trained to run the process before startup. The initial startup procedures and normal operating procedures need to be fully evaluated as part of the pre-startup review to assure a safe transfer into the normal operating mode for meeting the process parameters.

For existing processes that have been shutdown for turnaround, or modification, etc., the employer must assure that any changes other than "replacement in kind" made to the process during shutdown go through the management of change procedures. P&ID's will need to be updated as necessary, as well as operating procedures and instructions. If the changes made to the process during shutdown are significant and impact the training program, then operating personnel as well as employees engaged in routine and nonroutine work in the process area may need some refresher or additional training in light of the changes. Any incident investigation recommendations, compliance audits or PHA recommendations need to be reviewed as well to see what impacts they may have on the process before beginning the startup.

ANALYSIS OF THE REGULATION

(1) Design Specifications

There are two ways in which the correct implementation of design specifications can be checked. The PSSR team members can carry out spot-checks of the installed piping and equipment, and compare it with the piping lists and equipment data sheets. The second way is to make sure that a system of turnover packages (which are discussed below) has been implemented and followed.

(2) Procedures

The PSSR should check that operating procedures for any new operation have been written down. The standard also mentions maintenance procedures here, even though they are not identified elsewhere in the regulation as a separate activity, in the way that operating procedures are.

(3) New/Modified Facilities

The regulation requires that new facilities conduct a PHA. Therefore, the PSSR should check that the PHA was in fact carried out, and that its recommendations were either resolved or implemented.

During the pressure of construction, there is sometimes a tendency to postpone some of the PHA recommendations for "further study" as a convenient way of putting them off because there is insufficient time and/or money to implement them. The PSSR team members should carefully check any recommendations that were postponed, and satisfy themselves that such postponements do not jeopardize the safety of the plant.

LINKAGE OF PSSR WITH OTHER ELEMENTS

Since a PSSR occurs at the conclusion of a project, it has usually been preceded by a number of other PSM activities. These typically include:

1. Management of Change
2. Mechanical Integrity Reports (which includes Process Safety Information)
3. PHA's
4. Procedures and Training

A PSSR ensures that all these were carried out correctly.

MANAGEMENT OF CHANGE

Management of Change (MOC) is described in Chapter 12. The PSSR should check that all "not-in-kind" changes were put through the MOC system properly. Also, the PSSR should try to find covert changes, i.e. those changes which slipped through and which therefore never entered the MOC system. One way that the PSSR team can address this is to review all maintenance work orders and construction change orders. If any are found that are deemed to be more than just changes "in-kind", they should be identified, and put into the MOC system.

EQUIPMENT INSPECTION REPORTS

Inspection reports, which are an integral part of the Process Safety Information and Mechanical Integrity programs, should be included in the PSSR because they indicate where changes have been made, and so they serve as a means of cross-checking the Management of Change system.

PHA's AND OTHER SAFETY ANALYSES

PHA recommendations generate requests for change. The PSSR team should check that these changes were implemented properly, as discussed above.

PROCEDURES AND TRAINING

Those changes that do not change the actual operation — for example new values for Safe Upper and Lower Limits — still require a PSSR review. Not just process safety information, but operating procedures, training and contractor management should all be checked.

PSSR CHECKLIST

A PSSR can be organized and conducted like a PHA Checklist analysis. A representative PSSR checklist is shown in Table 9-1. This can be used as the basis for developing questions appropriate to the plant being analyzed.

Table 9-1

Representative PSSR Checklist

Engineering

1. Check that vessels are constructed according to vessel specification details.
2. Check that equipment is constructed and installed as per design.
3. Check that proper installation practices were followed.
4. Check that piping is installed as per the P&ID's.
5. Check that all process safety information has been updated. This includes relief valves, instruments and vent systems.

Operations

1. Check that all waste streams can be handled properly.
2. Check that all off-spec product can be handled.
3. Check that there is sufficient capacity to handle all materials in an emergency.
4. Check that Hot Work, Confined Space Entry, Lockout/Tagout and Work Permit procedures are being followed.
5. Verify that all operating procedures and checklists are up to date.
6. Verify that all operating and maintenance training is complete.
7. Check that all relief valves and rupture disks are present, and that all block valves in relief systems are locked open.
8. Check that all vessels have been closed and secured properly.
9. Check that all tools and materials have been removed.
10. Review emergency procedures, including changes to upper and lower safe operating limits.
11. Check that all emergency response personnel have been trained.

Process Hazards Analysis

1. Verify that a PHA was carried out.
2. Verify that findings were addressed properly.
3. Ensure that all personnel had access to the findings of the PHA.

Maintenance

1. Verify that maintenance procedures have been updated.
2. Verify that maintenance training has been provided.
3. Verify that all required maintenance materials are available.

Process Safety Information

1. Verify that all process safety information has been updated.
2. Verify that the P&ID's are accurate and up to date.
3. Verify that MSDS are up to date.
4. Verify that the process safety information is available to all workers, as needed.

TURNOVER PACKAGES

Turnover packages are often used as a means of efficiently transferring "care, custody and control" of a new facility from the construction to the operation department. Therefore, these packages can be used as a means of organizing the PSSR program. The use of turnover packages improves both efficiency and safety. Efficiency is improved because the process of turning over the system to operations can start very early on, thus saving time. Safety is improved because there is a clear definition of boundaries, with a corresponding understanding of who is responsible for what. Whenever the operations department is about to assume responsibility for a turnover package, a PSSR can be carried out on the contents of that package.

Turnover packages are used for new plants, and sometimes when extensive changes have been made to existing plants. Each package represents a discrete, self-contained section of the plant, such as the instrument air header or the boiler feed water system. When construction has finished all of its work on that system, it turns it over to operations. Once operations accepts the package, they assume full responsibility for all the equipment contained within it. By dividing a large facility this way, the operations department can start the testing and commissioning on the packages that have been turned over. It is not necessary to wait for the entire facility to be complete before commissioning can start.

When construction transfers custody of a turnover package to operations, they are saying that *everything* to do with that system is mechanically complete. This includes instruments, painting, insulation and all civil and structural work. The system should also have been blown free of trash before being handed over. Turnover packages can be contractually important. By transferring "care, custody and control" from construction to operations, the responsibility for fixing problems after that time no longer belongs to the construction team.

A turnover package is prepared using a master set of P&ID's. The turnover coordinator identifies each system and then defines it in detail, usually using a colored highlighter pen on the P&ID. Generally, each package will represent a single functional entity, such as a cooling water header or a reactor. If the turnover packages are prepared early enough, they can be identified on the P&ID itself as part of the general system of line and equipment markings.

An alternative method of turning over the plant is geographical. The plant is divided into sections, and all the piping and items within each section are completed and handed over to the operations department. The drawback to the geographical method is that it does not take into account the fact that the process systems, particularly those to do with utilities, stretch throughout the plant. Therefore, it may not be realistic to handover just part of say the cooling water system. However a geographical approach makes the handover of civil and structural work simpler.

As construction nears completion for each turnover system, the following work process is commonly followed:

1. All the items associated with each package are identified and listed, usually by function. So, for example, instruments, insulation and flange ratings will be listed separately This work can be done when construction is say 80% complete. Construction prepares a punch list for itself, which it will use to check out the work it is finishing. When they are ready, the Construction Department will punch out the plant using their list, and will fix any problems that they identify.

2. As completion approaches 100%, the operations department prepares its own punchlist for that package. When construction states that they are ready to turn it over, the Operations Department will punch out the unit; they do not rely on what was done by Construction.

3. Normally, there will still be a few items left incomplete. Operations may decide (during the PSSR review) that they can accept the plant in this condition because these missing items are not important to safety, and should not be allowed to slow down progress. They can be addressed once the plant is running.

4. Once the outstanding items on the worklist have been completed, the construction department will prepare a formal turnover letter. This letter is signed by both Construction and Operations to state that custody, care and control of that turnover package is now the

responsibility of operations. Once this letter has been signed, Construction can no longer work on anything defined within that turnover package without obtaining permission from Operations. From that point forward, there can be no guarantee that they do not contain hazardous or flammable chemicals. Hence the acceptance of care, custody and control means that standard operating safety systems, such as lockout/tagout will be in place.

POSTSTARTUP SAFETY REVIEWS

Once a facility has been commissioned and started, it is useful to carry out a safety review say 6 to 12 months after it has reached full operation. The objectives of such a review include:

- Ensure that all standards are being adhered to.

- Ensure that all outstanding safety recommendations have been properly closed out.

- Ensure that the operation is what was designed, and that any deviations have been properly managed through the Management of Change system.

- Identify any unanticipated operating problems that could affect safety.

- Feedback lessons learned to the research, development and engineering groups so that they can use the information in subsequent projects.

CONCLUSIONS

Prestartup and Restart Safety Reviews are an important part of process safety, yet are not always given the attention that they deserve. They provide a last chance for everyone associated with a project to make sure that no unsafe acts or conditions have slipped through before operations actually start. The temptation to rush this step must be avoided. The whole point is that people should be able to think about the safety of their facility before starting it.

Chapter 10

MECHANICAL INTEGRITY

INTRODUCTION

 A Mechanical Integrity (MI) program seeks to ensure that all equipment, piping, instrumentation, electrical systems and other physical items in a unit are designed, constructed and maintained to the appropriate standards, and the chance of failure of one of these systems is minimized, largely through the selection of inspection methods that are appropriate with regard to both technique and frequency for the systems being checked.

Mechanical Integrity can be a high profile topic because failures in this area are often so obvious, and they can lead to very serious accidents. Furthermore, Mechanical Integrity frequently receives a good deal of management attention because it can be very time-consuming and expensive to implement.

In most areas of process safety management each company has to develop standards appropriate its own operations. Mechanical Integrity tends to be an exception to this generalization. Many aspects of this topic are covered by codes and standards produced by a wide variety of agencies and regulatory bodies. The use of codes and standards does not, however, eliminate the need for judgment. The code writers recognize that every situation is unique, and that one rule cannot cover all situations.

The topic of mechanical integrity is a very broad one. In spite of the fact that its title contains the word "mechanical", it covers much more than mechanical engineering issues. One way of organizing "mechanical" integrity is to break it down according to traditional engineering and technical disciplines, as shown in Figure 10-1.

Figure 10-1

Some Elements Of Mechanical Integrity

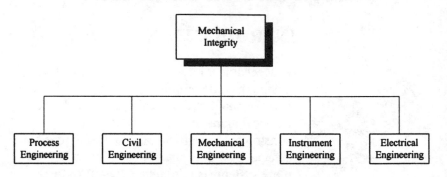

It is outside the scope of this book to cover all aspects of such a subject; indeed, it would require an entire library of engineering texts so to do. Therefore, the focus within this chapter is on developing a plan for the implementation and execution of a mechanical integrity program in a process facility.

THE OSHA REGULATION

(1) Application. Paragraphs (j)(2) through (j)(6) of this section apply to the following process equipment:

(i) Pressure vessels and storage tanks;

(ii) Piping systems (including piping components such as valves)

(iii) Relief and vent systems and devices;

(iv) Emergency shutdown systems;

(v) Controls (including monitoring devices and sensors, alarms, and interlocks) and,

(vi) Pumps.

(2) Written procedures. The employer shall establish and implement written procedures to maintain the on-going integrity of process equipment.

(3) Training for process maintenance activities. The employer shall train each employee involved in maintaining the on-going integrity of process equipment in an overview of that process and its hazards and in the procedures applicable to the employee's job tasks to assure that the employee can perform the job tasks in a safe manner.

(4) Inspection and testing.

(i) Inspections and tests shall be performed on process equipment.

(ii) Inspection and testing procedures shall follow recognized and generally accepted good engineering practices.

(iii) The frequency of inspections and tests of process equipment shall be consistent with applicable manufacturers' recommendations and good engineering practices, and more frequently if determined to be necessary by prior operating experience.

(iv) The employer shall document each inspection and test that has been performed on process equipment. The documentation shall identify the date of the inspection or test, the name of the person who performed the inspection or test, the serial number or other identifier of the equipment on which the inspection or test was performed, a description of the inspection or test performed, and the results of the inspection or test.

(5) Equipment deficiencies. The employer shall correct deficiencies in equipment that are outside acceptable limits (defined by the process safety information in paragraph (d)) before further use or in a safe and timely manner when necessary means are taken to assure safe operation.

(6) Quality assurance.

(i) In the construction of new plants and equipment, the employer shall assure that equipment as it is fabricated is suitable for the process application for which they will be used.

(ii) Appropriate checks and inspections shall be performed to assure that equipment is installed properly and consistent with design specifications and the manufacturer's instructions.

(iii) The employer shall assure that maintenance materials, spare parts and equipment are suitable for the process application for which they will be used.

OSHA GUIDANCE

Employers will need to review their maintenance programs and schedules to see if there are areas where "breakdown" maintenance is used rather than an on-going mechanical integrity program. Equipment used to process, store, or handle highly hazardous chemicals needs to be designed, constructed, installed and maintained to minimize the risk of releases of such chemicals. This requires that a mechanical integrity program be in place to assure the continued integrity of process equipment. Elements of a mechanical integrity program include the identification and categorization of equipment and instrumentation, inspections and tests, testing and inspection frequencies, development of maintenance procedures, training of maintenance personnel, the establishment of criteria for acceptable test results, documentation of test and inspection results, and documentation of manufacturer recommendations as to meantime to failure for equipment and instrumentation.

The first line of defense an employer has available is to operate and maintain the process as designed, and to keep the chemicals contained. This line of defense is backed up by the next line of defense which is the controlled release of chemicals through venting to scrubbers or flares, or to surge or overflow tanks which are designed to receive such chemicals, etc. These lines of defense are the primary lines of defense or means to prevent unwanted releases. The secondary lines of defense would include fixed fire protection systems like sprinklers, water spray, or deluge systems, monitor guns, etc., dikes, designed drainage systems, and other systems which would control or mitigate hazardous chemicals once an unwanted release occurs. These primary and secondary lines of defense are what the mechanical integrity program needs to protect and strengthen these primary and secondary lines of defenses where appropriate.

The first step of an effective mechanical integrity program is to compile and categorize a list of process equipment and instrumentation for inclusion in the program. This list would include pressure vessels, storage tanks, process piping, relief and vent systems, fire protection system components, emergency shutdown systems and alarms and interlocks and pumps. For the categorization of instrumentation and the listed equipment the employer would prioritize which pieces of equipment require closer scrutiny than others. Meantime to failure of various instrumentation and equipment parts would be known from the manufacturers data or the employer's experience with the parts, which would then influence the inspection and testing frequency and associated procedures. Also, applicable codes and standards such as the National Board Inspection Code, or those from the American Society for Testing and Material, American Petroleum Institute, National Fire Protection Association, American National Standards Institute, American Society of Mechanical Engineers, and other groups, provide information to help establish an effective testing and inspection frequency, as well as appropriate methodologies.

The applicable codes and standards provide criteria for external inspections for such items as foundation and supports, anchor bolts, concrete or steel supports, guy wires, nozzles and sprinklers, pipe hangers, grounding connections, protective coatings and insulation, and external metal surfaces of piping and vessels, etc. These codes and standards also provide information on methodologies for internal inspection, and a frequency formula based on the corrosion rate of the materials of construction. Also, erosion both internal and external needs to be considered along with corrosion effects for piping and valves. Where the corrosion rate is not known, a maximum inspection frequency is recommended, and methods of developing the corrosion rate are available in the codes. Internal inspections need to cover items such as vessel shell, bottom and head; metallic linings; nonmetallic linings; thickness measurements for vessels and piping; inspection for erosion, corrosion, cracking and bulges; internal equipment like trays, baffles, sensors and screens for erosion, corrosion or cracking and other deficiencies. Some of these inspections may be performed by state or local government inspectors under state and local statutes. However, each employer needs to develop procedures to ensure that tests and inspections are conducted properly and that consistency is maintained even where different employees may be involved. Appropriate training is to be provided to maintenance personnel to ensure that they understand the preventive maintenance program procedures, safe practices, and the proper use and application of special equipment or unique tools that may be required. This training is part of the overall training program called for in the standard.

A quality assurance system is needed to help ensure that the proper materials of construction are used, that fabrication and inspection procedures are proper, and that installation procedures recognize field installation concerns. The quality assurance program is an essential part of the mechanical integrity program and will help to maintain the primary and secondary lines of defense that have been designed into the process to prevent unwanted chemical releases or those which control or mitigate a release. "As built" drawings, together with certifications of coded vessels and other equipment, and materials of construction need to be verified and retained in the quality assurance documentation. Equipment installation jobs need to be properly inspected in the field for use of proper materials and procedures and to assure that qualified craftsmen are used to do the job. The use of appropriate gaskets, packing, bolts, valves, lubricants and welding rods need to be verified in the field. Also, procedures for installation of safety devices need to be verified, such as the torque on the bolts on ruptured disc installations, uniform torque on flange bolts, proper installation of pump seals, etc. If the quality of parts is a problem, it may be appropriate to conduct audits of the equipment supplier's facilities to better assure proper purchases of required equipment which is suitable for its intended service. Any changes in equipment that may become necessary will need to go through the management of change procedures.

ANALYSIS OF THE REGULATION

A brief overview of the regulation is provided here. Further detail is provided in the remainder of this chapter.

(1) Application

This section describes the equipment to which the standard applies. The list covers not just equipment and piping, but also instrumentation. It is assumed here that the word "Pumps" in paragraph (vi) covers all rotating equipment.

(2) Written Procedures

The development of procedures is a formidable task. Chapter 6 describes the development of operating procedures. One part of that Chapter discusses some of the issues to do with the writing of procedures for maintenance personnel.

(3) Training

This covers more than just maintenance training. All employees need to be aware of mechanical integrity issues. For example, Chapter 13 (Incident Investigation) describes an accident that was caused by the installation of an incorrect gasket. This involved an operator and a warehouse worker. Both of these needed training in mechanical integrity as it applied to their work.

(4) Inspection and Testing

Inspection and testing are the heart of a mechanical integrity program. Guidance as to how and when inspection and testing should be done is based on manufacturer recommendations, engineering practice and operating experience. As with all other elements of PSM, thorough documentation is needed.

One issue to watch for with regards to inspection is that the inspection activity may itself introduce an uncontrolled change into the system that could lead to an accident taking place.

(5) Deficiencies

Any identified deficiencies must be addressed. This means that a budget should be prepared to take care of those items that will require attention following the Mechanical Integrity inspections.

(6) Quality Assurance

In the context of process safety, quality assurance is more concerned with problems that lead to accidents rather than product quality in the customer-satisfaction sense. However, there is generally a close relationship between the two.

MANAGING MECHANICAL INTEGRITY

A Mechanical Integrity program can be developed using the following four steps:

1. Set Up An Information Base.
2. Prioritize The Risks.
3. Implement An Inspection Program.
4. Take Appropriate Corrective Actions.

1. SET UP AN INFORMATION BASE

The first step in any Mechanical Integrity program is to establish the information base. This will be an important part of the Process Safety Information program (Chapter 4.) The equipment information base will consist of a list of all equipment items (including instruments), the key operating parameters for each item, and its upper and lower design and operating limits for pressure and temperature. For mechanical equipment, this information will be usually be structured in the form of the data sheet that was developed when the item was designed and constructed.

2. PRIORITIZE THE RISKS

As with all aspects of process safety, it is necessary to sort and rank the information so that the critical items can be handled first. Hence, some information regarding the predicted frequency and consequences of failure is needed. The first step is to determine consequences of failure. Sanders has divided these into three categories for piping and equipment,

- *Priority 1* includes all those items that contain toxic or flammable materials. If there were to be a loss of containment from these vessels, a serious accident could ensue.

- *Priority 2* includes items that would not normally cause a serious accident should they fail. This would generally include equipment that is normally empty, and most utility systems.

- *Priority 3* covers those items where a failure would not cause a serious incident.

The information needed to determine these priorities will come from PHA's, expert judgment, incident histories and an assessment of industry requirements.

3. IMPLEMENT AN INSPECTION PROGRAM

Having set up the data base, the next step is to develop an inspection program that will provide information as to the actual condition of the equipment and instrumentation. This can be compared with the design conditions; any differences can then be judged according to the level of risk that they represent. The frequency and thoroughness of the inspection methods for each piece of equipment will depend on the level of risk associated with its failure, as described in the previous section. This program should include record-keeping and reliability assessments.

4. TAKE CORRECTIVE ACTIONS

The final step is to take corrective actions based on the results of the inspection program and a risk analysis of its results.

MECHANICAL ENGINEERING

Figure 10-1 divided Mechanical Integrity into different engineering disciplines. Usually the most important of these is the area of mechanical engineering, which, for the purposes of Mechanical Integrity, can itself be divided into the following areas.

1. Pressure vessels
2. Storage tanks
3. Piping and valves
4. Rotating equipment
5. Reciprocating equipment
6. Vent and relief systems

For all items, the following procedure should be followed.

1. Identify the affected equipment.
2. Establish the design basis/requirements for each equipment item.
3. Identify the inspection/test requirements for each equipment item.
4. Establish the frequency of inspection/tests.
5. Determine the acceptance criteria for acceptance/tests.
6. Establish the basis for procedures to alter or repair.
7. Establish proper installation procedures for new equipment.
8. Ensure that the maintenance materials and spare parts are availabel for all equipment items.
9. Ensure that all spare equipment and replacement parts are properly stored.
10. Develop procedures and training programs for all of the above.

Corrosion is one of the most common causes of equipment failure. Corrosion can be either uniform or localized. Uniform corrosion occurs evenly over a large metal surface and can involve large metal loss, hence the equipment involved may lose a considerable amount of mechanical strength. Localized corrosion can be seen in pitting and crevice attacks. It can lead to the creation of stress corrosion points.

Uniform corrosion can usually be predicted and monitored. Therefore, even though the ultimate consequence associated with this type of failure can be high, it unlikely to be a serious problem in practice because corrective action can be taken well before the problem becomes serious. In the case of localized corrosion, it is more difficult to monitor and predict consequences, hence it is usually a higher risk problem.

The different types of equipment are discussed in the following sections. In each case, widely used test and repair methods have been quoted. Naturally, these will be modified to reflect the plant's specific needs, and to conform to local regulations and standards.

PRESSURE VESSELS

Pressure vessels operate with internal pressures that are either above or below atmospheric pressure. They have a design pressure that represents the ultimate pressure that the vessel can withstand. Operating pressures should never reach the design pressure. Before they do so, either the cause of the high pressure (say an out-of-control reaction) should have been stopped, or a pressure relieving device should have activated in order to vent the high pressure fluids within the vessel.

Because of the potential danger associated with the rupture of pressure vessels, there is a large number of codes and standards to be followed. Generally, the operator of the equipment is provided with a Maximum Allowable Working Pressure (MAWP.) This is the maximum pressure at which an item of equipment can be operated, and is set by code. Because temperature affects the strength of a vessel (higher temperatures make the metal yield more easily), the MAWP has an associated temperature. If the temperature of the vessel is higher than what is specified, then the MAWP value is too high for safety.

The effect of high temperature on equipment strength can be very serious. For example, the design pressure for a certain type of pipe may be 150 psig at a temperature of 600°F. At 1000°F, the same piece of pipe will fail at just 20 psig. On the other hand, at 100°F, it may be able to handle nearly 300 psig. Hence, when temperatures are changing, the nominal pressure rating can be very misleading. (In this context, metal temperature refers to the average metal temperature through its entire depth.) Although low temperatures generally enhance metal strength (and so raise MAWP), very low temperatures may cause sudden and catastrophic embrittlement. This can be a serious problem is cryogenic services. For example, carbon steel equipment and pipe is liable to fail, even when there is no load on it, if its temperature falls to about -20°C. This can occur, for example, if a cryogenic liquids such as liquid air enter a carbon steel flare header.

If the MAWP is exceeded, the following effects are likely to be observed. At 1.5 to 2 times the MAWP, the vessel wall is close to yield. The vessel or pipe wall may be slightly distorted, but any leaks are most likely to occur at gaskets. At 2 to 4 times MAWP, there will probably be distortion of the vessel, and it can be assumed that gaskets will blow out. Above 4 times MAWP, there will be gross distortion of the vessel and/or pipes, and there may be a catastrophic total failure.

Although the MAWP should never be exceeded during normal operation, it may be acceptable for the operating pressure to go above the MAWP for brief periods of time, say during an emergency situation. However, following such an excursion, the vessel should be checked by qualified vessel expert before it is put back into service.

A cause of sudden and catastrophic failure is chemical embrittlement. This occurs when the wrong chemical enters the vessel or pipe, and causes it to fail in a very short period of time. Examples of this include the effect of caustic on stainless steels and hydrogen on various types of steel. Chemical embrittlement is particularly serious because it can happen to new equipment as easily as old, and because it may give little or no warning that it is about to occur.

There are many testing and inspection methods that can be used for pressure vessels. Some of the latest methods are discussed by Kenney. Representative test and repair methods for pressure vessels are shown in Table 10-1. (In Tables 10-1 through 10-6, various professional societies and groups such as AIChE, ASME and API are referred to. Frequently, OSHA will accept their guidance with respect to interpreting a regulation. However, this cannot be guaranteed in every case.)

Table 10-1

Methods For Pressure Vessels

Program Element	Program Requirements
Equipment Identification	Maintain a data base of all pressure vessels.
Design Basis/Requirements	Design per ASME Code Exceptions to be approved by MOC
Inspection/Test Procedures	Inspection API-510 Code Test ASNT-TC-1A
Inspection/Test Frequencies	API-510. Adjust according to operating experience
Inspection/Test Acceptance Criteria	ASME Code and National Board Inspection Code (NSIC) used to determine if repairs are needed
Inspection/Test Documentation	Maintain records
Repair and Alteration Procedures	API-510

Generally, equipment and piping is designed to a code such as ASME Section VIII Division I or B31.3. These codes consist of a set of rules that are based on historical performance, and that typically provide safety margins from 1.5 to 4. Alternatively, equipment can be designed by analysis. A technique such as finite element analysis is used to determine the actual stresses and strains at each point in the process, including potentially troublesome areas, such as piping elbows. The use of analytical methods means that it is not necessary to use large safety margins indiscriminately.

A problem that often arises on older plants is that there is no documentation to show that a particular pressure vessel was designed and constructed according to any established standard or code. As Hudson points out, the argument that a vessel has been in service for many years without problems is not satisfactory because it may already be at or over its anticipated life-span. Therefore, vessels such as these have to inspected very carefully, and then calculations have to be performed to show that it is still within the requirements.

STORAGE TANKS

Unlike pressure vessels, storage tanks are not designed to handle either high pressure (or vacuum conditions.) Typically, the tank is either open to the atmosphere, or to a part of the system (such as a flare header) that is guaranteed to be at atmospheric pressure.

Representative test and repair methods for storage tanks are shown in Table 10-2.

Table 10-2

Test Methods For Storage Tanks

Program Element	Program Requirements
Equipment Identification	Maintain a data base of all storage tanks.
Design Basis/Requirements	API: 12D,12F, 650 or 620
Inspection/Test Procedures	Inspection API-570. Test ASNT-TC-1A
Inspection/Test Frequencies	API-653 - adjusted according to operating experience
Inspection/Test Acceptance Criteria	API: 12D,12F, 650 or 620 to determine if repairs are needed
Inspection/Test Documentation	Maintain records
Repair and Alteration Procedures	API-653

PIPING AND VALVES

Many of the general comments made with respect to pressure vessels apply equally to piping. After all, a pipe can have a larger diameter and contain a higher pressure than many pressure vessels.

For refineries, the Piping Inspection Code usually followed is API 570. ASME B31.3 is also widely used. Items covered in these codes include the following:

- Inspector qualifications
- Inspection frequencies
- Inspection methods

- Evaluation of inspection results
- Repairs

Table 10-3

Inspection Methods for Piping and Valves

Program Element	Program Requirements
Equipment Identification	Maintain a data base of all piping and valves
Design Basis/Requirements	ASME/ANSI B31.1, B31.3
Inspection/Test Procedures	Inspection API-570. Test ASNT-TC-1A
Inspection/Test Frequencies	Inspection API-570 and operating experience
Inspection/Test Acceptance Criteria	ASME/ANSI B31.1, B31.3
Inspection/Test Documentation	Maintain records
Repair and Alteration Procedures	API-570

ROTATING EQUIPMENT

Equipment, particularly rotating equipment, is subject to dynamic loads. There are usually three types of failure, as described below.

CASING FAILURE

Such a failure is normally very serious because:

- It can lead to a major release of process chemicals.
- The casing often serves as a support for the complete equipment assembly.
- As it fails, the casing may send fragments of metal flying through the air — these can injure people and seriously damage other equipment.

Failure modes for casing includes overpressure, mechanical stress, and catastrophic reaction with the process materials. Although casing failures are likely to be extremely serious, they are also quite rare.

FAILURE OF THE ROTATING ELEMENT

If the internal rotating or reciprocating element within the equipment item fails, there is likely to be a major process upset. The cause will usually be either maintenance/construction problems or some type of interaction with the process materials. This type of failure is not likely to be serious in and of itself in terms of safety. However, it could have a serious impact on other parts of the unit.

SEAL FAILURES

Seal failures are the most common type of failure associated with process equipment. Since seals are a barrier between the process and the atmosphere, such failures are often likely to lead to environmental problems. Whether they become a safety problem depends on factors such as the flammabiliy/toxicity of the process fluid, the size of the leak and its proximity to where people are working.

A common cause of seal failures of centrifugal pumps is blocking in the pump while it is still running (or starting up the pump with the discharge block valve closed.) Although, this scenario is not usually hazardous, it may turn out to be so. Cooling water pumps, for example, have been known to explode when blocked in while running. The water in the casing is heated to the point where it starts to boil, the seal leaks but not enough to relieve the rapidly rising pressure, and so the casing ruptures.

Table 10-4

Inspection Methods For Rotating Equipment

Program Element	Program Requirements
EQUIPMENT IDENTIFICATION	Maintain a data base of all equipment.
Design Basis/Requirements	ANSI/API Codes
Inspection/Test Procedures	Manufacturer's recommendations
Inspection/Test Frequencies	Manufacturer's recommendations
Inspection/Test Acceptance Criteria	Manufacturer's recommendations. Company standards
Inspection/Test Documentation	Maintain records
Repair and Alteration Procedures	Manufacturer's recommendations. Company standards

VENT AND RELIEF SYSTEMS

Vent and relief systems are an essential component of virtually all plant safety systems. Whenever a vessel can be blocked in there is always the chance of it being over-pressured. Typical causes of over-pressure include:

- *External fire*. The heat from the fire causes the pressure inside the vessel to rise.

- *Chemical reaction*. A chemical reaction inside the vessel might cause its pressure to rise.

- *Connection to high pressure source.* If a high pressure system is connected to the vessel, then there is always a chance that the vessel will be over-pressured by the system.

Safety relief valves (SRV's) provide protection against an over-pressure situation. Thermal relief valves are found in services where a system, such as a long length of pipe, is liquid full and is subject to the heat of the sun. The rise in pressure caused by the heat is relieved by the thermal relief valve.

All relief valves must be inspected on a regular basis. Indeed, for some relief valves, such as those in boiler service, such inspections are mandated by law. Generally, the relief valve is removed from the vessel that it is protecting and tested in a workshop. The test will ensure that the relief valve opens at the specified pressure. Ideally, the test should also establish that the flow of gas is sufficiently high. During the test, an inspection should be carried out to check for corrosion, physical damage and any other problems that could affect the integrity of the relief valve. Sanders and Wood report the following testing frequency for a large petrochemical facility:

- 30% annually
- 60% every two years
- 10% on an individual basis

When a relief valve is removed for testing, the vessel that it was protecting no longer has that safeguard against over-pressure. Ideally, the plant will be shut down and depressured. However, in many facilities block valves are installed immediately below the relief valve. The block valve is closed, and the relief valve removed. During the time that the vessel is operating in this manner, it is necessary to have some other means of protecting the vessel. One way is to have an operator stand by a block valve that can be opened and that will depressure the vessel to a vent or flare. If he receives instructions from the control room, he will immediately open this valve. (Although this practice is followed in some locations, many facilities would not allow it on the grounds that it is much too risky.)

Whenever a valve or a rupture disk is placed below a relief valve, it is important to make sure that the pressure drop across it has been incorporated into the relief valve design calculations. Also, there is well-known problem to do with pinhole leaks in rupture disks below relief valves. If such a leak occurs, then the pressure on the downstream side of the disk will be roughly equal to the process pressure. Therefore, the process pressure required to rupture the disk may be almost twice as high as expected by the designer. Moreover, the pinhole may allow for the development of problems that the rupture disk was meant to avert (such as polymer build up on a relief valve.) Because of the danger associated with pinhole leaks, many facilities will install a pressure gauge in the spool piece between the rupture disk and the relief valve.

One particular problem to watch for is a build up of polymer on the inside of the relief valve. This can functionally disable the relief valve. If this is a chronic problem, it can be addressed by placing a rupture disk below the relief valve. Any polymer will stick to the bottom of the disk. Then, if there is a high pressure situation, the rupture disk will rupture and the gases will cause the (clean) relief valve to lift.

Relief valves and rupture disks are not the only form of overpressure protection device. In low pressure systems, particularly those which do not contain toxic or highly flammable materials, it may be sufficient to place hatches or doors on the system. If the pressure becomes too high, the hatch or door flies open. There are, however, a number of issues to watch for with such systems, including:

- Ensuring that the hatch or door is properly restrained (often with a hinge) so that it does not fly through the air, and

- Ensuring that, when the pressure returns to normal, the door or hatch does not close too quickly. If it were to do so, a vacuum condition might be created.

Table 10-5

Inspection Methods For Vent And Relief Systems

Program Element	Program Requirements
EQUIPMENT IDENTIFICATION	Maintain a data base of all vents, relief valves, rupture disks and flame arrestors
Design Basis/Requirements	ASME, API-530, RP2000, NFPA
Inspection/Test Procedures	NBIC (National Board Inspection Code), API-510, operating experience
Inspection/Test Frequencies	Operating experience
Inspection/Test Acceptance Criteria	Manufacturer's recommendations. Company standards
Inspection/Test Documentation	Maintain records
Repair and Alteration Procedures	Manufacturer's recommendations. Company standards

SAFETY INSTRUMENTATION

Safety instrumentation identifies and controls unsafe conditions before they proceed too far. It includes Alarms and Interlocks. The normal plant instrumentation is designed to keep the plant in a safe condition. However, it also has to control the plant for normal operations. Safety instrumentation is dedicated just to keeping the system safe. It also typically has a higher level of reliability than normal control instrumentation, and is tested much more often. Furthermore, on-line proof testing should be included for high-integrity devices such as these.

Because one of the hazards that may arise is that the normal control instrumentation fails to operate correctly, the safety instrumentation should be a completely separate system (which is one reason that safety instrumentation should have a higher reliability — it should be much more trustworthy than the control instrumentation.)

Table 10-6

Inspection Methods for Safety Instruments

Program Element	Program Requirements
Equipment Identification	Maintain a data base of all safety instrumentation
Design Basis/Requirements	ANSI, NEC
Inspection/Test Procedures	In-line or in-vessel instruments according to piping and valve requirements.
Inspection/Test Frequencies	Operating experience. Minimum of once per annum.
Inspection/Test Acceptance Criteria	In-line or in-vessel instruments according to piping and valve requirements.
Inspection/Test Documentation	Maintain records
Repair and Alteration Procedures	Manufacturer's instructions

CONCLUSIONS

Mechanical Integrity is a resource-intensive part of process safety. It requires input from a wide range of engineering disciplines, and covers many activities. As with other elements of process safety, Mechanical Integrity is probably best managed in a top-down mode, i.e. the initial program should cover the whole facility in a relatively low level of detail. Then, as problems are identified (both from the first phase of inspections and from other elements such as PHA's) the individual elements of the Mechanical Integrity program can be developed in greater detail.

Chapter 11

HOT WORK

INTRODUCTION

 Hot Work is that work that involves the use of flames, such as with welding or cutting torches, or that creates sparks such as when metal is being cut. There are four principal dangers from hot work.

1. Hot work provides a source of ignition that could ignite a vapor release from another location.

2. The person doing the hot work may cut through the wrong line or burn a hole in a piece of equipment. This could lead to the release of hazardous chemicals or cause a fire or explosion.

3. The person doing the work can be burned by the equipment that they are using.

4. In confined spaces, hot work can create a harmful atmosphere, leading to the possibility of a worker being overcome by fumes.

Because of the hazards associated with hot work, it is necessary to have a thorough and reliable system for controlling this type of work. Authority to perform hot work falls under the broader topic of "Energy Control Procedures" or "Permits to Work" — a topic which also covers items such as vessel entry, electrical isolation and the digging of trenches. Although OSHA's process safety standard confines itself to hot work, it is the overall topic of Energy Control Procedures that is discussed in this chapter. The OSHA standard itself consists primarily of references to other regulations in this area.

The basic intent of the standards in this area is to ensure that workers are not exposed to sources of high energy, particularly when doing maintenance or construction. These sources of high energy can take various forms, including toxic chemicals, heat, electricity, moving machinery, hydraulics, pneumatic equipment, falling objects, springs and coils, and falls from equipment. In all cases, it is essential that the workers do not come into direct contact with such sources of energy.

Because the hazards associated with this type of work are usually well understood, most companies already have in place a Permit to Work system that is functional and safe. If they did not, they would already be having too many accidents. Consequently, this element in the process safety regulation has not led to many changes in the way that companies operate.

THE OSHA REGULATION

(1) The employer shall issue a hot work permit for hot work operations conducted on or near a covered process.

(2) The permit shall document that the fire prevention and protection requirements in 29 CFR 1910.252(a) have been implemented prior to beginning the hot work operations; it shall indicate the dates authorized for hot work; and identify the object on which hot work is to be performed. The permit shall be kept on file until completion of the hot work operation.

OSHA GUIDANCE

Non routine work which is conducted in process areas needs to be controlled by the employer in a consistent manner. The hazards identified involving the work that is to be accomplished must be communicated to those doing the work, but also to those operating personnel whose work could affect the safety of the process. A work authorization notice or permit must have a procedure that describes the steps the maintenance supervisor, contractor representative or other person needs to follow to obtain the necessary clearance to get the job started. The work authorization procedures need to reference and coordinate, as applicable, lockout/tagout procedures, line breaking procedures, confined space entry procedures and hot work authorizations. This procedure also needs to provide clear steps to follow once the job is completed in order to provide closure for those that need to know the job is now completed and equipment can be returned to normal.

ANALYSIS OF THE REGULATION

The PSM standard for Hot Work is basically a cross-reference to 29 CFR 1910.252(a), which is the provision of the Welding, Cutting and Brazing Standards dealing with fire prevention and protection. The pertinent sections of it are quoted below. (The full standard is about three times longer than what is quoted here.)

The basic precautions for fire prevention in welding or cutting work are:

(i) Fire hazards. If the object to be welded or cut cannot readily be moved, all movable fire hazards in the vicinity shall be taken to a safe place.

(ii) Guards. If the object to be welded or cut cannot be moved and if all the fire hazards cannot be removed, then guards shall be used to confine the heat, sparks, and slag, and to protect the immovable fire hazards.

(iii) Restrictions. If the requirements stated in [the above paragraphs] cannot be followed then welding and cutting shall not be performed.

(2) Special precautions. When the nature of the work to be performed falls within the scope of paragraph (a)(1)(ii) of this section certain additional precautions may be necessary:

(i) Combustible material. Wherever there are floor openings or cracks in the flooring that cannot be closed, precautions shall be taken so that no readily combustible materials on the floor below will be exposed to sparks which might drop through the floor. The same precautions shall be observed with regard to cracks or holes in walls, open doorways and open or broken windows.

(ii) Fire extinguishers. Suitable fire extinguishing equipment shall be maintained in a state of readiness for instant use. Such equipment may consist of pails of water, buckets of sand, hose or portable extinguishers depending upon the nature and quantity of the combustible material exposed.

(iii) Fire watch. (A) Fire watchers shall be required whenever welding or cutting is performed in locations where other than a minor fire might develop, or any of the following conditions exist:

{1} Appreciable combustible material, in building construction or contents, closer than 35 feet (10.7 m) to the point of operation.

{2} Appreciable combustibles are more than 35 feet (10.7 m) away but are easily ignited by sparks.

{3} Wall or floor openings within a 35-foot (10.7 m) radius expose combustible material in adjacent areas including concealed spaces in walls or floors.

{4} Combustible materials are adjacent to the opposite side of metal partitions, walls, ceilings, or roofs and are likely to be ignited by conduction or radiation.

(B) Fire watchers shall have fire extinguishing equipment readily available and be trained in its use. They shall be familiar with facilities for sounding an alarm in the event of a fire. They shall watch for fires in all exposed areas, try to extinguish them only when obviously within the capacity of the equipment available, or otherwise sound the alarm. A fire watch shall be maintained for at least a half hour after completion of welding or cutting operations to detect and extinguish possible smoldering fires.

(iv) Authorization. Before cutting or welding is permitted, the area shall be inspected by the individual responsible for authorizing cutting and welding operations. He shall designate precautions to be followed in granting authorization to proceed preferably in the form of a written permit.

(v) Floors. Where combustible materials such as paper clippings, wood shavings, or textile fibers are on the floor, the floor shall be swept clean for a radius of 35 feet (10.7 m). Combustible floors shall be kept wet, covered with damp sand, or protected by fire-resistant shields. Where floors have been wet down, personnel operating arc welding or cutting equipment shall be protected from possible shock.

(vi) Prohibited areas. Cutting or welding shall not be permitted in the following situations:

(A) In areas not authorized by management.
(B) In sprinklered buildings while such protection is impaired.
(C) In the presence of explosive atmospheres (mixtures of flammable gases, vapors, liquids, or dusts with air), or explosive atmospheres that may develop inside uncleaned or improperly prepared tanks or equipment which have previously contained such materials, or that may develop in areas with an accumulation of combustible dusts.
(D) In areas near the storage of large quantities of exposed, readily ignitable materials such as bulk sulfur, baled paper, or cotton.

(vii) Relocation of combustibles. Where practicable, all combustibles shall be relocated at least 35 feet (10.7 m) from the work site. Where relocation is impracticable, combustibles shall be protected with flameproofed covers or otherwise shielded with metal or asbestos guards or curtains.

(viii) Ducts. Ducts and conveyor systems that might carry sparks to distant combustibles shall be suitably protected or shut down.

(ix) Combustible walls. Where cutting or welding is done near walls, partitions, ceiling or roof of combustible construction, fire-resistant shields or guards shall be provided to prevent ignition.

(x) Noncombustible walls. If welding is to be done on a metal wall, partition, ceiling or roof, precautions shall be taken to prevent ignition of combustibles on the other side, due to conduction or radiation, preferably by relocating combustibles. Where combustibles are not relocated, a fire watch on the opposite side from the work shall be provided.

(xi) Combustible cover. Welding shall not be attempted on a metal partition, wall, ceiling or roof having a combustible covering nor on walls or partitions of combustible sandwich-type panel construction.

(xii) Pipes. Cutting or welding on pipes or other metal in contact with combustible walls, partitions, ceilings or roofs shall not be undertaken if the work is close enough to cause ignition by conduction.

(xiii) Management. Management shall recognize its responsibility for the safe usage of cutting and welding equipment on its property and:

(A) Based on fire potentials of plant facilities, establish areas for cutting and welding, and establish procedures for cutting and welding, in other areas.
(B) Designate an individual responsible for authorizing cutting and welding operations in areas not specifically designed for such processes.
(C) Insist that cutters or welders and their supervisors are suitably trained in the safe operation of their equipment and the safe use of the process.
(D) Advise all contractors about flammable materials or hazardous conditions of which they may not be aware.

The OSHA PSM standard also refers to the NFPA (National Fire Protection Association) Standard for Fire Prevention in Use of Cutting and Welding Processes (51B, 1962.) One of its primary concerns is with the isolation of hot work from flammable materials. The following steps must be followed:

(1) Remove the object to be welded or cut to a safe place, i.e. away from the flammable materials.

(2) If (1) cannot be done, remove the flammable materials to a safe place.

(3) If (2) cannot be done, place guards around the work to prevent heat, sparks and slag from reaching the fire hazard.

(4) If (3) cannot be done, the work shall not be performed.

While the work is being performed, fire extinguishers must be in place, and a fire watch shall be maintained.

TYPES OF ENERGY CONTROL PROCEDURE

There are four basic types of Energy Control Procedure. They are illustrated for isolating a vessel from a line in Figure 11-1.

<u>Figure 11-1</u>

<u>Illustration Of Four Levels Of Security</u>

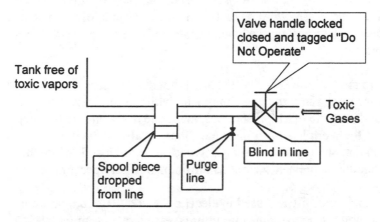

REMOVE THE HAZARD

The first and best of controlling energy is simply to remove the hazard. For example, if it is necessary to clean out the inside of a storage tank, it is best to completely drain the tank of chemicals, and then purge it with air until the atmosphere in the tank is safe to breathe. It may be possible to avoid doing this by having the workers wear special breathing equipment that allows the work to be done while the hazardous chemicals are still in the tank. However, this is more hazardous then cleaning out the tank before people enter it.

POSITIVE ISOLATION

If the hazard cannot be removed, a high level of security can be achieved by providing some form of positive isolation. In the case of piping, such isolation can be obtained by removing a section of the pipe (a spool piece), or by inserting a metal blind rated at the piping pressure.

LOCKOUT

The third type of Energy Control Procedure is one that prevents the worker from coming into contact with the source of high energy, using a device that is locked in place (hence the phrase "lockout"), with only authorized persons having the key needed to remove the locks. For example, if a valve is closed, and then secured with a padlock and chain, workers on the non-process side of that valve are protected. (Removing the handle of the valve also serves to "lock" it closed.)

In spite of the security that a lockout system provides, there are still hazards. First, the valve may leak while in the closed position. Second, in spite of all precautions, someone may remove the lock before the work has been finished. There is also a chance of confusion, such that the wrong valve is chained closed, while the valve that should have been secured is left in its normal operating state.

Lockout is always used in electrical work. The junction box containing the appropriate circuit breakers have an arm which is thrown into the "off" position and then padlocked in place. This prevents the system being energized while a worker is in contact with it.

ADMINISTRATIVE CONTROLS/TAGOUT

The fourth kind of Energy Control Procedure, and the one that provides the least protection, is the use of a warning sign, informing workers that they are adjacent to a hazardous situation, and providing instructions about what must be done to make sure that they are not exposed to the source of high energy. These warning signs often take the forms of tags with wires that are twisted on to the item of concern; hence, the phrase "tagout." Tagout does not provide good protection because it relies completely on administrative procedures to ensure the safety of the workers; there is no form of physical protection such as is obtained from removing the hazard, or locking it out.

Tags are almost always used with the other methods. For example, a lock on a valve handle will have a tag on it providing information such as who has authority to remove the lock and what the work order number for this job is.

LOCKOUT/TAGOUT

The topics of lockout and tagout have already been discussed at the start of this chapter. Additional information is provided here.

Lockout/tagout is a system used to protect workers from exposure to sources of high energy, and to ensure that confined spaces are not entered until they are completely safe. Lockout/Tagout is a procedural process; it creates systems for the control of sources of high energy.

A device is said to be locked out when it has been securely isolated against the inadvertent introduction of energy, and when a lock and key are used to make sure that it cannot be re-energized without following proper procedures. Lockout is very important, and failure to follow its requirements has led to many accidents. For example, Rekus states that about 100 fatalities per annum in the United States are due to failure to implement or properly execute Lockout/Tagout systems.

Tagout refers to the use of tags on the items that have been secured. The tag should clearly indicate which position the energy-isolating device needs to be in so as to be in the safe or off situation. Tags are always applied to locked-out devices.

In addition to the standards already quoted, further guidance on lockout/tagout is provided in the PSM standard, 29 CFR 1910.147, and in the booklet OSHA 3120: Control of Hazardous Energy (Lockout/Tagout.) The following points should be noted with respect to this standard.

Lockout/tagout standards do not cover cord and plug connections for electrical equipment because the hazard can be controlled simply by unplugging the equipment (however, the plug must be properly controlled so that no one inadvertently puts it back in the socket.)

Hot tap operations, in which hot work is carried out on a line or a piece of equipment while fluid is flowing through it are excluded from the lockout/tagout provisions. This type of operation is inherently risky. However, for those companies that conduct hot tap work, there are safety measures that can be followed in this area.

Minor maintenance work that is routinely carried out, and that is not construed to be hazardous is excluded. Regardless of these exceptions, lockout/tagout applies when (a) an employee must remove or bypass machine guards or other safety devices, (b) an employee is required to place his or her body in contact with, or in the operating area of the point of operation of a piece of equipment.

Lockout must always be used unless there is no physical means for its application. New equipment and machines must always be supplied with lockout capabilities.

In general, the originator of a lock or tag is the only person authorized to remove it. If there is a shift change, workers on the first crew must sign off the job and remove their locks, and workers on the second crew must sign on to the job and add their locks. (Lock boxes and master locks are usually used, as discussed on page 279.)

For a lockout system to work effectively, there must be a clear administrative procedure to (a) determine what items need to be isolated, and (b) to identify who has the authority to add and to remove locks.

The permit system should also define who has control of what, and for how long. Items to consider include the following:

- Identification of persons responsible for preparing the work location. Usually this will be the operators. They are charged with removing all sources of hazardous chemicals and energy.

- Identification of persons who will check that the clearance has taken place properly. This person may be responsible for application of a gas detector, or other safety instrument.

- Time that the job is allowed to take, after which it must be shut down, and the permit re-issued.

- Identification of who has keys to which locks, and where the master lockout plan is located.

- The sign off and turnover procedure once the job is finished.

- Identification of the person(s) responsible for checking that the job was completed as required by the permit.

ENERGY CONTROL PROCEDURES

The major steps in an Energy Control Procedure are described below.

PLAN THE TASK

The first step obviously is to determine what the task is, and how it is to be done. This step also means identifying and listing the affected pieces of equipment, and determining what impact it is likely to have on the overall operation. For a large, continuously operating unit, an important part of the plan is to determine what other parts of the facility will have to be shut down, and to determine how those items that are to be left running are to operate during this particular maintenance task. The planning involves determining how each affected unit is to be isolated (see below.)

It is very important at this stage to determine if the maintenance procedure falls under Management of Change.

SHUT DOWN THE EQUIPMENT

The next part of the plan is to determine how the affected unit(s) are to be shut down. This will probably involve the use of Temporary

Operating Procedures (see Chapter 6), and an explanation as to how the units still in service are to operate. Items to watch for include:

- Sudden changes in utilities consumption
- Inventories of intermediate chemicals
- Changes in flowrates

ISOLATE THE AFFECTED UNITS

The next step is to isolate the affected units. Every piece of equipment is "attached" to sources of high energy. In the case of a pump, for example, it could be attached to most of the following:

- *Rotating Energy.* The driver, drive shaft and impeller can all turn. It is important that they be secured from inadvertent movement (even in the motor has been de-energized.)
- *Electrical Energy.* If the pump has an electrically-driven motor, it is vital that this be properly isolated.
- *Heat Energy.* If the pump is driven by a steam turbine, or if there is steam tracing around it, it is important to ensure that the steam — and the associated steam condensate system — are properly isolated.
- *Chemical Energy.* If the pump normally handles hazardous chemicals that are toxic or a health hazard, it has to be properly cleared of these.
- *Flammable/Explosive Energy.* If the pump handles hydrocarbons, or other materials that could ignite, they have to be cleared, often using an inert gas such as nitrogen.
- *Potential Energy.* If the pump is not located at grade, it may be possible for a person to fall off it (even if it is at grade, there may be a pit below it.)

Each of these sources of energy must be isolated, and an energy control policy for each implemented. The policy for each may be different. For example, it is feasible to lockout electrical energy, but the only way of preventing people from falling, is to put barrier tape and other warning devices in the dangerous areas — in other words, to "tagout" potential fall areas.

One of the dangers associated with this step is that some sources of energy may be connected to the item being worked on, but no one has identified it. Major process lines and electrical systems will be identified, but some of the minor flushing lines, for example, may be overlooked, even though they contain hazardous chemicals. This provides yet another reason for making sure that the plant has to keep up-to-date P&ID's.

RELEASE THE STORED ENERGY

Once all sources of high energy have been identified, isolated and locked, the residual energy within the system has to be removed. Usually this means clearing the equipment to be worked on, and its line up to the isolation points, of residual chemicals and flammables. This is often an uncertain time; no matter how conscientiously this is carried out, there is always a chance that something will be overlooked. This is why the First Break policy, as described below, is so important.

A common example of this occurs with tanks and other storage vessels. They may slowly corrode at or near their base, resulting in the formation of pockets within the metal wall, or between layers of metal. Hazardous or flammable material will fill these pockets. Then, when the tank or vessel is cleared for hot work, a welder will put his torch over one of these pockets, leading to a fire.

APPLY THE LOCKOUT/TAGOUT

The next step is to apply the lockout/tagout device to prevent the high energy source from being reactivated or reintroduced.

When more than one person is working on a job, it is necessary for the Energy Control Procedure to define differing levels of responsibility: a Group Lockout/Tagout procedure is needed, one in which each person involved can apply their own lock, and only they can remove it. One way of doing this is to use a Masterlock system. The details will vary from company to company, but the following process is representative.

1. The lead person on the job locks closed each valve, switch and other device that is used to isolate the high energy source. Only the lead has a key to these locks.

2. He labels the locks with the appropriate work permit information, and then attaches a lock box to each location.

3. He places the keys in the box, closes it and then locks it with a master lock (which is often a distinct color such as red.)

4. Every worker who is to work on that job attaches his or her lock to the lock box so that it cannot be opened. They record what they have done on the work permit.

5. If a worker leaves the job, he or she removes their lock and signs off on the work permit. At the end of the job, each worker removes their individual locks and signs off the job. Each person must satisfy themselves that the job has, from their point of view, been returned to a safe condition before they remove their personal lock.

6. The lead then removes the master lock and takes the key to the valve from the box.

7. He then closes the work permit, and unlocks the valve or switch.

The person who is in overall control of the project is responsible for developing the overall master plan for the lockout/tagout process. For a complex job involving multiple valves and switches, they can prepare a chart using a P&ID showing where locks, tags and blinds are to be located.

- Equipment description
- Location of the work
- Work description
- Name of person(s) requesting the work
- Management of change clearance
- Sources of high energy normally present
- Isolation points
- Lock/tag applications at each isolation point
- Shutdown procedures
- Methods for removing stored/residual energy
- First break procedures
- Identification of all personal tags and locks

- Sign off procedures for job completion
- Sign off procedures for putting equipment back in service

VERIFY

Once the system has been prepared for work, and the locks have been applied, the system must be verified. If it is a motor that is being worked on, for example, the area should be cleared in case the isolation procedures fails, and then an attempt should be made to run the motor. If it does, then clearly the isolation mechanism has failed. If it is a valve that is being locked closed, the safety lead should try to open it after the locks and chains have been applied.

FIRST BREAK POLICY

If a line is being broken, or a cover is being taken off a vessel, first-break policies can be used. The first break is when the workers are actually exposed to the system for the first time. This is a dangerous point because, if the previous procedures have not been carried out properly, there may be an unexpected release. Therefore, the persons involved in making the first break should wear a higher level than normal of PPE.

START THE WORK

Once the first break has occurred, and assuming that everything is in order, the workers can start the maintenance task.

With regard to vessel entry, the following guidelines provide the basis for a policy.

1. Loosen bolting on the manway and remove all but four bolts. The remaining bolts should be positioned at 12, 3, 6, and 9 o'clock.

2. The last bolts on the manway should be loosened and carefully spread open to ensure that there is no pressure trapped in the vessel.

3. When it is confirmed that there is no residual pressure in the vessel, the four bolts can be removed, and the manway taken off.

4. When the manway cover is off, the manway opening should be covered to prevent unauthorized entry.

5. Blowers or air-movers may be installed on the manways to help cool and ventilate the vessel before it is entered.

6. There must always be a manway attendant equipped with an atmospheric monitoring instrument located at the manway when people are in the vessel.

7. People entering the vessel must always wear safety harnesses.

RETURNING THE EQUIPMENT TO SERVICE

The process just described is repeated in reverse when the equipment is being readied for return to service. Although the process itself may not be difficult, it is important to maintain careful vigilance, because this is a point at which people may be tempted to skip steps in order to get the unit back into full operation.

For a large job, the return of equipment to service may be part of the Restart Safety Review system (see page 233.)

LEVELS OF ISOLATION FOR VALVES

When isolating a piece of equipment from a hazardous process, there are various levels of security that can be applied. Some of the more commonly are listed below. The direction is in a direction from less to more secure, i.e. Level 7 is better than Level 1. (The discussions assume that valves should be closed; in some cases it may be necessary to keep a valve open. However, the same principles apply.)

LEVEL 1 —CLOSED VALVE

The lowest level of security is a closed valve. It is almost never acceptable as a means of isolation, except in the most benign services. There are two problems with using a closed valve to isolate a worker from a hazardous chemical. First, it is very easy for someone else to come by and open the valve. They may have mistaken the valve for one that is in service, or they may not have known that a job was in progress. Regardless of the reason, the valve can be easily opened by mistake. The second drawback to using just a closed valve is that it can be expected to leak. Indeed, except in the case of specially designed valves, it can be assumed that it will leak; very few valves provide a perfect seal.

LEVEL 2 — TAGGED VALVE

If it is not possible to do more than close a valve, some measure of security can be obtained by putting a tag on it. The tag will state who put the tag on the valve, and who has the authority to remove it.

LEVEL 3 — CAR SEALED CLOSED VALVE

A car sealed closed valve is one that is tagged in some manner to avoid the first of the problems identified above: inadvertent opening of the valve. The traditional form of car seal has been a wire loop with a tag attached to it — rather like a stronger version of a luggage tag. Modern car seals often use plastic ties that are similar to garbage bag ties (except that they are bigger.) The person placing the car seal writes instructions on the tag as to who has authority to remove the tag.

In most cases, car seals do not provide enough security to protect someone working on a piece of equipment. Nor do they prevent the problem of leakage through a valve that is closed but not leak proof. However, they do reduce the chance of someone opening the valve inadvertently. However, because the car seal has very little physical strength, it is very easy to break if someone is in a hurry, and they feel that the car seal should be removed.

For these reasons, car seals are usually used in non-safety situations. In the case of a blending operation, for example, one line to the blender may be car sealed to make sure that the product is not contaminated with what is in that line. However, were the wrong material to be blended, there would be no safety consequences.

284 Process Safety Management

LEVEL 4 — LOCKED CLOSED VALVE

A locked valve provides much more security. Generally, the person placing the device puts a chain around the valve and valve handle, and then padlocks the chain in such a manner that the handle cannot be turned. Once more, this does not eliminate problems to do with inadvertent leakage, but it does almost eliminate inadvertent valve movement.

LEVEL 5 — DOUBLE BLOCK AND BLEED/VENT VALVE SYSTEM

If a line has to be isolated on a regular basis, it is often convenient to provide a double block and bleed system, as shown in Figure 11-2. There are two block valves in the main process line. Between them is a bleed / vent (the bleed points down if the hazardous chemical is liquid; the vent points up if the chemical is vapor.) The closed block valves may also be blinded (see below.)

The bleed / vent valve represents a low pressure spot in the system. Therefore, if there is a release from the high pressure pipe, it will be diverted through the bleed, and it will not enter the equipment being worked on. If the discharge from the bleed / vent line could itself be hazardous, it can be routed to a safe location, such as a flare header.

Figure 11-2

Illustration of Double Block and Bleed

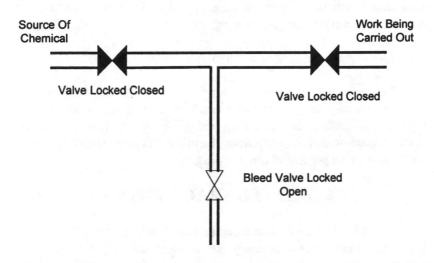

Source Of
Chemical

Work Being
Carried Out

Valve Locked Closed

Valve Locked Closed

Bleed Valve Locked
Open

LEVEL 6 —LINE BLIND(S)

Line blinding, which is often done in conjunction with the other methods, requires that a flat piece of metal be bolted to the flange at the end of the exposed line. The blind should be specified to take full line pressure plus the appropriate safety margins. In order to check that a blind is in place, it usually has a short handle which juts out from between the flange faces.

If a flange is to be routinely blinded, a spectacle blind (also known as a figure of eight blind) can be installed. This looks like the number eight, with one section closed, and the other open. When the flange is broken and the line cleared, the blind can be rotated around the one bolt that is left in the flange.

There are two problems to watch for with blinds. The first is to make sure that it is specified correctly. There have been numerous incidents where blinds have buckled when exposed to full system pressure. The second problem concerns accidentally forgetting to remove blinds when the work is over. During a large turnaround, there may be hundreds of blinds installed in the system. As the maintenance work closes out, there is usually considerable pressure to get the plant up and running as quickly as possible. It is very easy to overlook a blind, then, when the plant is started, it will be found that the blinded line cannot be operated. The removal of the blind at this time can be hazardous, particularly as there is likely to be a lot of management pressure to keep the start-up moving.

LEVEL 7 — PHYSICAL DISCONNECTION AND BLINDING OFF

The safest way of isolating a work site from hazardous chemicals is to remove a spool piece (a short length of pipe) from the system, and then to blind both of the remaining ends. This means that there is no route through which the hazardous chemical can enter the area where the maintenance personnel are working.

CONFINED SPACE ENTRY

A confined space is defined as an area whose primary function does not cover human occupancy, which has restricted entry and exit and which may contain hazards. Normally, a person should not enter a confined space until it has been securely isolated (using procedures such as those described above) and until the oxygen level is above 19.5 – 20.9% and less than 23.5% by volume. Also, there should be no toxic gases present, and the flammability of the atmosphere should be less than 10% of the Lower Explosive and Flammable Limits.

Generally, confined space entry procedures contain three major sections. The first is to do with the person performing the work. It covers items such as safety harnesses, ventilation, communication with support systems and breathing gear. The second part of procedure is concerned with the immediate support system — sometimes called the buddy system. It describes how the person doing the work communicates with his buddy, how he or she can be pulled out of the confined space in the event of an accident, and how additional help can be obtained. The third part of the procedures is concerned with backup and emergency response systems, i.e. what will happen if an accident does occur.

A person is said to have entered a confined space if they break the plane of the confined space portal with his or her face. Therefore, it is not permissible, for example, to take a breath and to put one's head into a vessel for a quick look. Proper entry procedures must be followed for even a quick "look-see."

If the conditions in a confined space change at any time, the entry permit must be re-issued. Furthermore, it may be policy to reissue the permit every 24 hours. The person who is in attendance at the manway has the following duties:

- Maintain an accurate count of the number of people in the vessel
- Maintain constant two-way communication with all of the occupants
- Provide standby assistance to occupants entering the confined space
- Instruct occupants to leave the confined space if any conditions change such that the entry permit is invalidated
- Monitor for changes that could lead to the occupants to be exposed to hazardous conditions
- Remain at the entry point at all times

A copy of the entry permit, and the restrictions that apply, should be posted at each entry point to the confined space.

CONCLUSIONS

The topic of hot work does not usually require a high degree of attention when setting up process safety programs because it is an area that is almost always well under control. Nevertheless, the development of a PSM program does provide an excellent opportunity to review this important topic, particularly with regard to the management and training of contract workers.

Chapter 12

MANAGEMENT OF CHANGE

INTRODUCTION

In Chapter 1, it was pointed out that all accidents result from uncontrolled change. Therefore, the control of change is at the heart of any PSM program. This is why the Management Of Change (MOC) system is so important. Its purpose is to make sure that all changes are properly identified and controlled before they are implemented.

One issue that management must face is that MOC appears to contradict other management principles, such as empowerment, in which individuals are given more freedom to make decisions on their own authority. MOC is a control system; as such it places constraints on individual freedom, and requires that everyone work within an established system or bureaucracy. A balance needs to be struck — management must establish control systems, but the system must not be too cumbersome or people will not use it, or they will try to bypass it (possibly by declaring all changes to be "emergency" changes.)

Implementing MOC can be difficult because it can cut across the "way things are around here" and "you've got to do it this way if you want to get anything done" attitudes. For example, operators may decide to store hard-to-get items such as special gaskets and fittings in their lockers. Their justification is that is cuts through the red tape. If they need one of these items in a hurry, they have it on hand, and can get the job done right away. Although such practices may indeed improve performance most of the time, the lack of control implicit in such a situation is disquieting. Eventually, there is likely to be a serious accident as a result of this action such that all the previous savings will be instantaneously canceled out.

THE OSHA REGULATION

(1) The employer shall establish and implement written procedures to manage changes (except for "replacements in kind") to process chemicals, technology, equipment, and procedures; and, changes to facilities that affect a covered process.

(2) The procedures shall assure that the following considerations are addressed prior to any change:

(i) The technical basis for the proposed change;

(ii) Impact of change on safety and health;

(iii) Modifications to operating procedures;

(iv) Necessary time period for the change; and,

(v) Authorization requirements for the proposed change.

(3) Employees involved in operating a process and maintenance and contract employees whose job tasks will be affected by a change in the process shall be informed of, and trained in, the change prior to start-up of the process or affected part of the process.

(4) If a change covered by this paragraph results in a change in the process safety information required by paragraph (d), such information shall be updated accordingly.

(5) If a change covered by this paragraph results in a change in the operating procedures or practices required by paragraph (f), such procedures or practices shall be updated accordingly.

OSHA GUIDANCE

To properly manage changes to process chemicals, technology, equipment and facilities, one must define what is meant by change. In this process safety management standard, change includes all modifications to equipment, procedures, raw materials and processing conditions other than "replacement in kind." These changes need to be properly managed by identifying and reviewing them prior to implementation of the change. For example, the operating procedures contain the operating parameters (pressure limits, temperature ranges, flow rates, etc.) and the importance of operating within these limits. While the operator must have the flexibility to maintain safe operation within the established parameters, any operation outside of these parameters requires review and approval by a written management of change procedure.

Management of change covers issues such as changes in process technology and changes to equipment and instrumentation. Changes in process technology can result from changes in production rates, raw materials, experimentation, equipment unavailability, new equipment, new product development, change in catalyst and changes in operating conditions to improve yield or quality. Equipment changes include among others change in materials of construction, equipment specifications, piping prearrangements, experimental equipment, computer program revisions and changes in alarms and interlocks. Employers need to establish means and methods to detect both technical changes and mechanical changes.

Temporary changes have caused a number of catastrophes over the years, and employers need to establish ways to detect temporary changes as well as those that are permanent. It is important that a time limit for temporary changes be established and monitored since, without control, these changes may tend to become permanent. Temporary changes are subject to the management of change provisions. In addition, the management of change procedures are used to insure that the equipment and procedures are returned to their original or designed conditions at the end of the temporary change. Proper documentation and review of these changes is invaluable in assuring that the safety and health considerations are being incorporated into the operating procedures and the process.

Employers may wish to develop a form or clearance sheet to facilitate the processing of changes through the management of change procedures. A typical change form may include a description and the purpose of the change, the technical basis for the change, safety and health considerations, documentation of changes for the operating procedures, maintenance procedures, inspection and testing, P&ID's, electrical classification, training and communications, pre-startup inspection, duration if a temporary change, approvals and authorization. Where the impact of the change is minor and well understood, a check list reviewed by an authorized person with proper communication to others who are affected may be sufficient. However, for a more complex or significant design change, a hazard evaluation procedure with approvals by operations, maintenance, and safety departments may be appropriate. Changes in documents such as P&ID's, raw materials, operating procedures, mechanical integrity programs, electrical classifications, etc., need to be noted so that these revisions can be made permanent when the drawings and procedure manuals are updated. Copies of process changes need to be kept in an accessible location to ensure that design changes are available to operating personnel as well as to PHA team members when a PHA is being done or one is being updated.

ANALYSIS OF THE REGULATION

(1) Written Procedures

Like all the elements of process safety, Management of Change procedures must be written down. This ensures consistency, and it means that they can be audited. This section of the standard introduces the phrase "replacement in kind." The basic idea is that if an item is being replaced by something that is absolutely identical to it, then a change has not occurred, so the MOC process does not need to be implemented.

In most cases, this is not a problem. But someone may make a change inadvertently. They believed that they were making replacing something with an identical copy, but they were not, or else they erroneously assumed that it was "close enough."

Sanders provides an example of an accident that took place in which a worker placed glycerin, rather than the normal liquid, in a bubbler that scrubbed a small flow of gas containing nitrous and nitric acids. The glycerin reacted with these gases to form nitroglycerin, which detonated two days later. In as much as the worker considered MOC at all, he presumably felt that the change he was making was no more than a Replacement In Kind.

Although it is usually clear when a change in equipment has occurred, there are some gray areas, such as the following.

1. An item, such as a valve, may be replaced by another item that is from another supplier, but that is equivalent to the original. In these cases, if there is any change in material, design, shape, size or design, then this is not a replacement in kind, and MOC needs to be carried out. It is likely that the second valve will not be identical in size to the original. In this case, the key would be 'design specifications.' If they are identical, then it can be argued that the valve is identical.

2. If an item is relocated, it has "changed."

3. If an item is to be upgraded in standard, then MOC must be performed. For example, if it has been found that a section of pipe was not of the correct material of construction, then the replacement pipe is subject to MOC, even though it is of a higher standard because the new material may create some other type of system failure.

4. Other changes that are not in-kind include: change in electrical classification, change in lubrication systems, change in instrument ranges and change in purpose for using a chemical.

(2) Making The Change

If the change is not "in kind", it should be properly analyzed for its impact on safety and health. For a large change, one that requires a full AFE (Authorization for Expenditure) and PHA studies, this type of analysis will usually be performed anyway. However, small, quick changes are more likely to be made with little or no analysis.

(3), (4), (5) Training/PSI/Operating Procedures

Many changes will result in a different way of operating the plant. Therefore, it is important as part of the MOC program, that all affected individuals are informed as to the change, and trained in the new mode of operation. Also, Process Safety Information and the Operating Procedures must be updated.

TYPES OF CHANGE

Most discussions about MOC are concerned with intentional or overt changes. However, at least three different types of change can be identified.

OVERT

When someone wishes to make a change in a system, they must be taught the rules of the MOC system. Then they will know which changes need to be highlighted and reviewed before they are implemented. Although this is fundamentally a matter of good judgment, there are guidelines that can be provided. The following types of change will almost certainly require that the MOC system be used.

Changes To Process Chemistry

If the change involves the introduction of a new chemical (or the removal of an old one) MOC will be required. New chemicals could create undesirable side reactions or change heats of reaction such that a hazardous situation is created. They may also require changes to the Emergency Response Plan and they may require that a health analysis needs to be carried out.

Changes In Materials Of Construction

Changes in materials of construction will require MOC approval. Issues to watch for include corrosion, erosion, stress cracks and high and low temperature capability.

Changes In Operating Conditions

Every important process parameter has associated with it a safe upper and lower limit. By definition, if conditions change such that those limits are exceeded, then MOC is required. This applies to intentional change, such as when management wants to increase production rates, as well as unintentional changes. If an operator sees that his system has gone outside these bounds, he knows that there is a safety issue, and that some type of action must be taken.

Changes In The Control System

Changes in control systems will usually require MOC. This includes changes in logic, set points (as defined by the safe upper and lower limits for the variable in question) and response parameters. Any changes to the emergency shutdown system and its associated interlocks will certainly need to go through the MOC process.

COVERT SUDDEN

Sometimes changes are made without anyone being aware of it. Therefore, by definition, the MOC will not identify them. In one refinery, for example, an operator was changing the gasket on a filter housing — a routine task. What he did not know was the that gasket was of the wrong specification. When he restarted the filter, the gasket failed and highly flammable liquid escaped. There was a fired heater nearby, so the liquid ignited. The resulting fire caused an immense amount of damage (fortunately no one was hurt.)

Changes such as these cannot be handled by the MOC system because no one is aware that a change has occurred.

COVERT GRADUAL

Even more difficult to spot are covert changes that take place gradually. On one chemical plant there was a set of large, liquid-filled reactors. A noble metal compound catalyst, which was in solution, was fed into the reactors. Over a period of years, the noble metal had slowly formed a metal lining that was steadily growing in thickness on the inside walls of the reactors. The process and economic consequences of this problem had been analyzed and accepted. However, no one had considered the civil engineering implications of this uncontrolled change: the fact that the reactors were slowly getting heavier, and possibly overloading their foundations.

A more common covert, gradual change concerns safety systems such as the firewater and flare headers. On many plants, more and more equipment is installed over the years, thus potentially overloading such systems. Since these are passive systems, there are no indications that they have become overloaded, and there is a danger that analyses of these systems will be put off until it is too late.

CRITICAL SAFETY SYSTEMS

If the change involves the temporary removal or bypass of a critical safety system, the implications of the move need to be examined very carefully. A common example is the removal of relief valves for routine bench testing. While the relief valve is removed, it is vital to have an adequate backup safety system.

An even more difficult system to work on is the flare header. Even when a plant is totally shut down, there are inventories of hazardous chemicals stored in tanks, which normally vent to the flare. Isolation of this header, therefore, means that alternative vent locations have to be found.

FORMS OF CHANGE

Most changes fall into one of three major groups: equipment, administrative systems or engineering. On most facilities, these changes will come either from Maintenance Work Requests or from Engineering Work Orders (CITGO.)

EQUIPMENT

Equipment changes tend to be the focus of most MOC systems, and they are usually well controlled. If accidents due to equipment changes do occur, it is often associated with small items.

ADMINISTRATIVE

Many changes require changes to the administrative systems, particularly operating procedures and training. This is something that may be overlooked during the MOC process.

ENGINEERING

The most widely quoted example to do with inadequate MOC is the accident that occurred in Flixborough, England in 1974. It led to 28 fatalities, many injuries and catastrophic economic loss. The root cause of the accident was the installation of a temporary pipe that had not been properly engineered, and that failed when put in service. (The reason that this modification was needed in the first place was that another uncontrolled equipment change had taken place, resulting in the need for the temporary pipe.)

Another example of doubt about whether a change has occurred occurs with training. If a new operator moves on to a unit, and this operator has extensive experience with similar plants but not with this particular unit, the question will be raised as to whether a formal change has occurred, or whether his existing "grandfathered" knowledge constitutes a change-in-kind.

Even more difficult is the situation where the people involved know that they are making a change, but they feel that it is in the best overall interest of the plant to over-ride the MOC system (often using the emergency MOC procedures) so that their project can get finished.

IMPLEMENTATION

SYSTEM DESIGN

The central problem that all MOC systems face is balancing the need for thoroughness with the problems of too much bureaucracy. The system has to be thorough, otherwise potentially hazardous changes will slip through. On the other hand, if the system becomes too cumbersome and slow, people will avoid using it.

The number of change in most plants is large. For example, it was reported at a 1996 Texas Chemical Council meeting that the number of changes that flow through the MOC system is typically around 250 per year for a medium-sized site (with say 140 employees) and up to 1,000 a year for a large site with say 2,000 employees. It was further reported that about 75% of the changes that enter the system can be regarded as moderate, i.e. they are not perceived as materially affecting the safety of the unit. Up to 10% are serious enough that they have to be reviewed with a checklist or some other type of PHA.

MOC can occur at various stages in a project. For example, an MOC may be performed on a conceptual design. It can then be repeated when details of the equipment design are available. Woodruff suggests that each change should have the following factors documented.

1. What is the change
2. Who is making the change
3. Timing
4. Reasons for the change
5. Expected benefits
6. Who is affected
7. Who will resist
8. Reasons for resistance
9. Other concerns

There are three elements to most MOC programs.

1. Review all changes, and highlight those that are not a "change-in-kind", i.e. the direct replacement of one item or system by another where there is no change in requirements of functionality.

2. For those changes that are not "in-kind", determine if the new item or system is hazardous, and, if it is, take the appropriate corrective action. In extreme cases, this may mean abandoning the change altogether on the grounds that there is no way of implementing it without incurring unacceptable risk.

3. Review the change as it is being implemented, and follow up to make sure that there were no inadvertent side-effects associated with the change. Also, ensure that those changes that are temporary are discontinued at or before their termination date.

Almost all MOC's will create a need to update the operating procedures because the procedures describe the man-machine interface. Generally, the new procedure should be written within 24 hours of the change being made, and operators should be trained and tested on the change within 3 days. Most MOC's will also require that the Process Safety Information base be updated. It is particularly important to make sure that the P&ID's have been updated because of their importance to the remainder of the PSM program.

All changes will require that operators and other affected individuals be trained and/or informed as to how the process is different from what it was before the change.

MAKING THE CHANGE

The process that most companies follow for MOC is usually quite straightforward (it has to be, otherwise it will not be used.) Generally, it follows the steps described below.

1. The person wanting to make the change fills out an MOC form. This explains what he or she wants to do, why the change is needed, and what the safety implications may be.

2. A small MOC committee made of representatives from different departments (operations, maintenance and engineering must be on the committee) reviews the change and either authorizes it or not.

3. The MOC committee decides whether any further analysis of the change is needed. For example, they may decide that a PHA is required. Moreover, they may choose to specify the level of PHA (HAZOP, What-If, etc.) depending on the nature of the proposed change. They will also make it clear that the appropriate changes need to be made to Operating Procedures, Training and Process Safety Information.

4. The committee may also decide to appoint someone to monitor the change as it is implemented, and to report on any unexpected problems.

5. Finally, and this is very important, the committee must make sure that the person who recommended the change be kept fully informed as to the status of his recommendation, particularly if the committee decides to turn it down.

FIELD CHANGES

Kenney (189) describes the Field Change Authorization system as used by his company. These are essentially changes that can be approved and executed by operations supervisors; there is no need for any time of external review from say one of the technical departments. In order to make sure that all field changes are correctly assigned — i.e. that they do not require further examination — there may be a Field Change Supervisor to assist with that decision.

Field changes must be reported to a central MOC committee, even if they have already been executed. This gives management to make sure that no unacceptable changes have been allowed to slip through.

EMERGENCY CHANGES

Inevitably, some changes will have to be implemented more quickly than the MOC system, no matter how streamlined it may, will permit. Therefore, a system for implementing emergency changes is needed. Generally, the justification for an emergency change is that someone in line authority, a shift supervisor for example, decides that the danger associated with doing nothing is greater than the danger associated with not having the change properly authorized. For example, a gasket may start to leak, allowing hazardous chemicals into the atmosphere. The warehouse may report that they do not have identical gaskets with which to replace the leaking item, but they do have others which are very similar. In this case, the operations and maintenance supervisors may decided to go ahead and make the change right away, even though it is not a Replacement In Kind.

The problem with having the capacity for making emergency changes is that the system can be abused, particularly if the normal MOC system is perceived as being cumbersome and slow. Therefore, regardless of the urgency with which the change has to be made, there must always be a system for providing the necessary authority for making such a change. Possibly the people with the authority to make the changes can be on call. On a large facility, it may be possible to have one person always on site who has the authority to make changes — this may be the emergency director, the person who is responsible for operating the unit in the event that an emergency is declared.

TEMPORARY CHANGES

Many changes will have a finite time for which they are valid. For example, temporary bypasses or the use of a tank to store a non-standard chemical are changes that will be stopped as soon as the reason for the change has been eliminated. Changes of this type require a built-in termination date. It is important that the MOC system have the capability of handling such situations by prompting the affected parties to tell them that the termination time has been reached.

DOCUMENTATION

Some of the documents that have to be checked and possibly updated after a change has been made include the following:

- P&ID's, including equipment lists and line lists
- Instrument loops, including alarm settings and interlock logic
- Operating procedures

- Operator training records
- Preventive maintenance procedures
- Isolation valve list
- Electrical one-line diagrams
- Blind lists
- Inventory levels
- Relief valve information

CONCLUSIONS

Management of Change is at the heart of process safety. If all accidents are caused by uncontrolled change, then the identification of changes (particularly those that are covert), and the correct handling of them from that point on is critical. The development of a change system is quite simple in principle. However, it requires a good deal of discipline to make sure that it is properly working, particularly in administrative areas such as operating procedures and training.

Chapter 13

INCIDENT INVESTIGATION

INTRODUCTION

Many elements of a PSM program, such as Process Hazards Analysis and Prestartup Safety Review, are directed toward anticipating problems, and then suggesting the corrective actions that can be taken to ensure that these problems do not actually occur. The limitation with such techniques is that, regardless of the quality and thoroughness of the analysis, the findings are theoretical; there is no assurance that the high probability and high consequence scenarios have actually been identified. This is why Incident Investigations are so important. They provide hard data as to how accidents really can happen. Moreover, incidents provide a wake-up call; they show that accidents can happen, even in the best-managed plants, and that safety can always be improved.

Accidents and incidents are different from one another. An incident is an event that has either caused harm or that has the potential to cause harm. (Incidents are analogous to hazards; a hazard represents no more than a *potential* for harm.) For example, if the pressure in a vessel rises above its MAWP (maximum allowable working pressure) there is the potential for an explosion. This is an incident. If the relief valve on the vessel does not work, and the pressure continues to rise causing the vessel to rupture, then the incident has become an accident. Similarly, a spill of gasoline is an incident. If the gasoline ignites and burns someone, then the incident has become an accident.

For an Incident Investigation system to be effective, it must analyze all incidents: including both accidents and near-misses. The reason for this is that there are so many more incidents (near misses) than there are accidents. For example, a representative ratio of accidents to incidents is shown in Table 13-1, which is based on work reported by Bradley for DuPont sites.

Table 13-1

Incident Ratios

Type of Incident/Accident	Number
At-Risk-Behavior	300,000
Near Misses	30,000
Recordable Injuries	300
Lost-Time Injuries	30
Fatality	1

The data of particular interest in Table 13-1 are Near Misses and Lost-Time Injuries. A Lost-Time Injury is always very serious, and its occurrence will undoubtedly lead to a formal incident investigation. Lost-Time injuries are also rare. Hence it is difficult to build up a useful knowledge base from these events. Near Miss incidents are much more common. Table 13-1 suggests that there may be 1000 Near Misses for each Lost-Time injury. Investigation of them can therefore provide a statistically significant data base which can be used for root cause analysis (discussed later in this chapter.)

In risk analysis terms, a Near Miss situation represents the occurrence of one, but not all, of the events entering an AND in a fault tree (discussed in Chapters 5 and 16.) If all the events entering an AND Gate occur, then the incident becomes and accident. This is shown in Figure 13-1 for the relief valve example just provided. If the pressure rises to a dangerous level (the initiating event) AND the relief valve does not operate (the safeguard fails), THEN the Top Event (vessel rupture) will occur. The incident of high pressure becomes the accident of vessel failure.

Figure 13-1

Incident/Accident Fault Tree

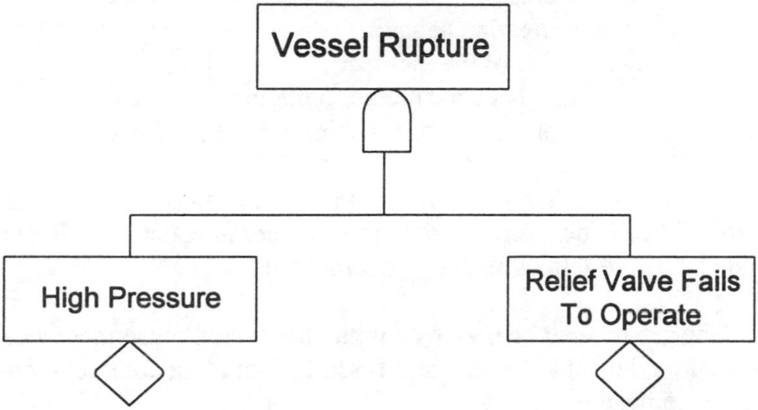

In any type of investigation, those leading the effort should take every step possible not to blame people for accidents and incidents. The purpose of the investigation is to identify management problems, not to find fault. However, if the investigations shows that one person or group of people did a superior job, then this should be highlighted and those who did the work should be recognized.

THE OSHA REGULATION

(1) The employer shall investigate each incident which resulted in, or could reasonably have resulted in a catastrophic release of a highly hazardous chemical in the workplace.

(2) An incident investigation shall be initiated as promptly as possible, but not later than 48 hours following the incident.

(3) An incident investigation team shall be established and consist of at least one person knowledgeable in the process involved, including a contract employee if the incident involved work of the contractor, and other persons with appropriate knowledge and experience to thoroughly investigate and analyze the incident.

(4) A report shall be prepared at the conclusion of the investigation which includes as a minimum:

(i)   Date of incident;
(ii)   Date investigation began;
(iii)   A description of the incident;
(iv)   The factors that contributed to the incident; and
(v)   Any recommendations resulting from the investigation.

(5) The employer shall establish a system to promptly address and resolve the incident report findings and recommendations. Resolutions and corrective actions shall be documented.

(6) The report shall be reviewed with all affected personnel whose job tasks are relevant to the incident findings including contract employees where applicable.

(7) Incident investigation reports shall be retained for five years.

OSHA GUIDANCE

Incident investigation is the process of identifying the underlying causes of incidents and implementing steps to prevent similar events from occurring. The intent of an incident investigation is for employers to learn from past experiences and thus avoid repeating past mistakes. The incidents for which OSHA expects employers to become aware and to investigate are the types of events which result in or could reasonably have resulted in a catastrophic release. Some of the events are sometimes referred to as "near misses," meaning that a serious consequence did not occur, but could have.

Employers need to develop in-house capability to investigate incidents that occur in their facilities. A team needs to be assembled by the employer and trained in the techniques of investigation including how to conduct interviews of witnesses, needed documentation and report writing. A multi-disciplinary team is better able to gather the facts of the event and to analyze them and develop plausible scenarios as to what happened, and why. Team members should be selected on the basis of their training, knowledge and ability to contribute to a team effort to fully investigate the incident. Employees in the process area where the incident occurred should be consulted, interviewed or made a member of the team. Their knowledge of the events form a significant set of facts about the incident which occurred. The report, its findings and recommendations are to be shared with those who can benefit from the information. The cooperation of employees is essential to an effective incident investigation. The focus of the investigation should be to obtain facts, and not to place blame. The team and the investigation process should clearly deal with all involved individuals in a fair, open and consistent manner.

ANALYSIS OF THE REGULATION

(1) Investigation

An employer must investigate incidents, as well as accidents, if such as release could have led to a catastrophic release. Therefore, the above discussion about investigating near misses is incorporated into the regulation.

(2) Timing

The investigation should be started within 48 hours at the latest. It does not permit time off for holidays or other scheduled breaks.

(3) Team

The investigation has to be carried out by a qualified team.

(4) Follow Up

There must be prompt follow up to the investigation. All persons involved shall be informed as to the findings.

REPORTING INCIDENTS

If the reporting of all incidents, including near misses, is to be encouraged, then the facility must have a culture where workers are willing to communicate problems. This means that they must know that they will not be reprimanded (except for fair cause.) Management must look upon incidents as a means of improving its systems, not as a way of finding fault with employees. This is difficult, particularly when someone *is* at fault, and disciplinary action has to be taken.

One way of building confidence in the system is to make sure that the employee who first reports the incident or near miss is kept fully informed as to the status of the report and management's follow-up. (This is yet another form of Employee Participation.) Also, it is important not to make incident reporting too much of a hassle or too bureaucratic. Otherwise people will avoid using it.

When a company develops an open attitude toward incident reporting, it may notice an increase in the number of recorded incidents. This does not mean that more incidents are actually taking place — it simply means that fewer mishaps are being swept under the rug and that people are reporting situations more freely.

Some companies restrict the reporting of incidents due to legal liability concerns. Therefore, each company needs to work with its own attorneys on developing means whereby information on accidents and near-misses can be disseminated freely without causing too much legal liability for the company.

PERFORMING AN INCIDENT INVESTIGATION

An Incident Investigation is similar to the audit process described in Chapter 15. It can be divided into the following steps:

1. Accident Response/Report the Incident
2. Determine the facts of the incident
3. Identify root causes
4. Issue a report and recommendations
5. Follow up

ACCIDENT RESPONSE/REPORT THE INCIDENT

If the incident actually resulted in an accident, the first step is to respond to the accident itself, and to reduce the amount of damage caused. In its book on this topic, CCPS describes some of the initial actions to be carried out. Additional information on this accident response is also provided in OSHA's 29 CFR 1910.120, the HAZWOPER standard. Accident response includes the following:

I. Rescue all injured personnel.

II. Provide medical treatment to all affected personnel.

III. Secure the site so that the accident cannot restart, or have effects elsewhere.

IV. Make sure that the environmental impact is minimized.

V. Inform the pertinent agencies.

VI. Collect and preserve physical evidence, using photographs as much as possible. This includes:

    A. Record as-found positions for all valves, instruments, switches
    B. Note residual liquid inventories, flame marks.
    C. Check all gaskets, relief valves and rupture disks.
    D. Collect all paper and electronic records. This includes strip charts, logs, logbooks and instruction books
    E. Catch samples for future analysis

VII. Gather all the Process Safety Information that might be needed. This can include:

    A. MSDS
    B. P&IDs
    C. Instrument loop data

        D.   Instrument and interlock set points (the widespread use of Distributed Control Systems, in which large quantities of operating data are recorded for extended periods of time by computers which are physically remote from the accident site, greatly assists this)

VIII.   Start the investigation. This is discussed in the following sections.

IX.   Clean up the site once all evidence has been collected.

X.   Repair and re-start (after the reasons for the incident have been determined and the appropriate actions taken to make sure that it does not happen again.

If the incident was a near miss that could have led to a serious accident, it will be investigated and reported at the same level as if it had been an actual accident.

DETERMINE THE FACTS OF THE INCIDENT

For a formal investigation, an Incident Investigation team will have to be formed. The size of this will depend on the seriousness of the incident. The team will normally consist of people who work in the plant. Outside specialists may be used if they have particular expertise. If the incident is very severe, or if it is felt that independent leadership is important (maybe due to pending litigation) the company may choose to recruit an outside consultant to lead the investigation.

The team leader should have attributes similar to those of a good PHA team leader. He or she should be very experienced in plant operations so that they can understand "the ways things really are", and can pick up on some of the unspoken communication that takes place. Following an incident, people may be reluctant to be completely candid. This is not to say that they are intentionally lying, but they may be reluctant to be completely candid. The leader should be able to read some of the unspoken signals so that he can direct his or her investigation in the correct direction.

The team members can be selected from the following:

- Operations person from another plant
- Process engineer
- Safety department
- Operations supervisor from affected plant
- Worker from unit affected

IDENTIFY ROOT CAUSES

The Incident Investigation must find out the systemic cause of the incident. If an operator made an error, for example, the root cause could be inadequate operating procedures or insufficient training. Hence, the final recommendation will probably have nothing to do with that particular worker; it will be concerned with improving procedures or training.

Root cause analysis can be pushed back almost indefinitely. If the training program is inadequate, there must be a cause for that, and so on. Therefore, it is important to provide a sensible framework in which root causes can be identified, without the company spending an inordinate amount of time and money on the analysis, and without coming up with findings that are very difficult to implement. One way of conducting a root cause analysis using the PSM elements is described later in this chapter. Fault Tree Analysis (*see* Chapter 16) can also be used.

ISSUE A REPORT AND RECOMMENDATIONS

All Incident Investigations will require that a report be written. Depending on the seriousness of the incident, there can be three levels to the report:

1. Preliminary
2. Interim
3. Final

(1) Preliminary Report

The preliminary report is written almost immediately, say within 24 hours of the incident, particularly if the incident is an actual accident (as distinct from a near miss) and regulatory agencies have to be notified right away. This report is likely to contain just a brief description of the incident, how much damage (safety, environmental and economic) was done, and what the emergency response team did in response to the incident.

(2) Interim Report

Once the Incident Investigation has been completed, an Interim Report can be issued. This will describe the incident in some detail, and provide at least some root cause analysis. It will also generate short-term recommendations.

(3) Final Report

The final report may be issued much later than the first two. There are two possible reasons for it taking more time. First, the writers of the report may be waiting on information from lab tests and other time-consuming activities. Second, the report will have to be very carefully reviewed within the company and by legal counsel for accuracy and for what it might say in terms of potential liability.

The final report should include a thorough root cause analysis, along with the recommendations arising from that analysis. As a minimum, the final report should include the following items:

- Date of Incident
- Date of Investigation
- Description of the Incident
- Factors that Contributed to the Incident
- Recommendations
- Review of the Report
- System to Address Findings
- Retention Period for the Report

There are three important aspects to follow-up. First, there needs to be an audit process to ensure that recommendations are implemented in a timely manner. This includes the short-term recommendations that can come out of the preliminary reports, along with longer-term recommendations from the final report.

The second part of follow-up is to make sure that all employees and other personnel who could have been affected by the incident are properly informed as to what happened. Details of the incident, and what was done to prevent it from recurring can be published in a plant newsletter, or other forum. If it is very important that the company be able to demonstrate that the incident was properly communicated, a bulletin concerning it can be included with the employees' pay envelopes, because this is one packet that will always be opened.

The guidance provided in Chapter 15 — Audits — concerning the way in which to structure and write a report of this nature are also pertinent to Incident Investigation.

Finally, there should always be an evaluation and critique of the Incident Investigation process itself.

ROOT CAUSE ANALYSIS

Root cause analysis is a vital part of Incident Investigation. The problems that are often faced in this area include how to structure the analysis, and how deep to make the analysis.

THE ELEMENTS OF PROCESS SAFETY

One way of organizing the work is to use the elements of PSM to provide a framework for the discussion. This is shown for ten of the elements of Process Safety Management in Figure 13-2. (Trade Secrets have been eliminated from the list because they are essentially legalistic. Also Process Hazards Analysis, Audits and Incident Investigation are eliminated because they are all investigative techniques that do not directly affect the occurrence of incidents.)

Figure 13-2

PSM Elements In Root Cause Analysis

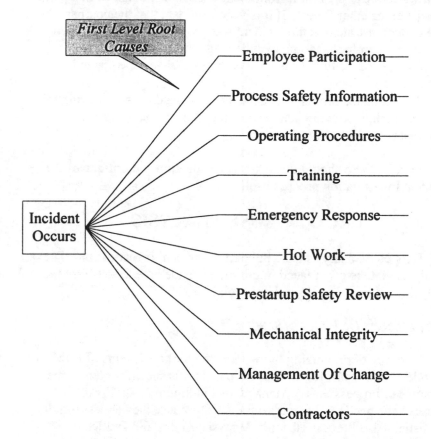

The Incident Investigation team works through each of the ten elements. Each can be analyzed by determining if there had been any deviations from design or operating intent. A PHA approach (*see* Chapter 5) can be used for this because basically a PHA is no more than a simulated Incident Investigation, in which accidents are postulated based on What-If questions that investigate the consequences of equipment items failing. Deviation guidewords can then be used to analyze the elements.

In the case of Operating Procedures, for example, the key words may be as follows:

1. *Too Many* Operating Procedures
2. *Not Enough* Operating Procedures
3. *Excessive Detail* in Operating Procedures
4. *Insufficient Detail* in Operating Procedures
5. *Other Than* Operating Procedures

Although this analysis may look a little strange to process engineers, it can be effective. Considering the points one by one.

Too Many Operating Procedures is not a common problem. But some companies do write so many procedures that eventually the Operating Manual becomes overwhelming, and hence it is ignored.

Not Enough Operating Procedures is an even more common problem. The incident being investigated may not have any type of procedure to cover it.

Excessive Detail is a real problem also. If an operator is familiar with an operation, he is not likely to bother reading the operating instructions if he feels that he has to wade through too much detail.

Insufficient Detail is fairly self-explanatory. Either there is no information at all about the situation in question, or the information provided is too general or too vague.

Other Than Operating Procedures refers to situations where the procedures deviate from the intent of providing instructions to the operators. Instead they may provide information that would better be included in the basic training program, say on how to start a centrifugal pump.

EXAMPLE

The idea of using PSM elements is illustrated in the following example (which is largely fictional, but is loosely based on an actual incident.)

An operator was changing a filter in a line containing a hazardous hydrocarbon with properties similar to that of gasoline. This was a routine operation which this operator had carried out many times. The top of the filter was bolted down on to the filter housing.

After changing the filter, the operator lined up feed to it using a valve adjacent to the filter. The gasket on the top of the filter failed, and a large quantity of hydrocarbon sprayed out, drenching the operator. There were four furnaces in the area, any one of which could ignite a fire at the leak. The operator was unable to stop the leak, so he quickly left the scene and discarded his saturated clothing. A minute or so later, the leak did ignite, and a bad fire ensued. Fortunately no one was hurt.

The preliminary investigation showed that the immediate cause of the accident was that the wrong type of gasket had been installed. The gasket that had been installed was very similar, but not identical to the correct gasket. Gaskets were supplied to operations by the warehouse. The preliminary investigation also showed that there had been no obvious cause of blame. For example, everyone concerned had carried out their job correctly, to the best of their knowledge.

Investigation of the accident in terms of the PSM elements identified the following problem areas.

Employee Participation

The warehouse personnel responsible for the proper storage and distribution of gaskets and other materials were contract workers with an inadequate knowledge of the plant's overall process safety program. They were unaware that what might appear to be a small change could lead to a serious accident. Therefore, even though they were at fault for issuing the wrong gasket, they had no knowledge of the potential consequences of their actions. Furthermore, management had never solicited the views of these workers as to what aspects of their work might lead to an error. For example, a change in the way that materials were stored or labeled might greatly reduce the chance of this type of incident from occurring again.

Process Safety Information

Problems with Process Safety Information were part of this accident. Specifically, the operator was not provided with the information needed to identify the right type of gasket.

Operating Procedures

Follow up showed that there was not an Operating Procedure for the gasket replacement activity. Nor were there any procedures concerning the selection of gaskets.

Training

This was a major factor in this incident. The operator was not trained to recognize gaskets. Nor were the warehouse personnel fully trained in their tasks.

Emergency Planning And Response

Once the incident occurred, the response of the emergency teams was excellent. This Incident Investigation showed that this element of the PSM standard was very well implemented. It is important to highlight successes such as this.

Mechanical Integrity

On first hearing of this incident, it would be easy to say that there was a problem with the Mechanical Integrity program. In fact, this is not the case. The specifications of equipment were never a problem.

In the event that mechanical integrity had been a problem, Bloch has identified the following as root causes.

1. Design
2. Assembly / Installation
3. Fabrication / Manufacturing
4. Material Defect
5. Maintenance Deficiency
6. Out-Of-Range Service Conditions

Contractors

At first, this element does not appear to be relevant to this example. All the people concerned (including those in the warehouse) were full-time employees. However, the gasket was supplied by a vendor. Whether or not the vendor is considered to be a contractor in terms of the standard, the principle is applicable: an outside company was involved in the accident. If the operating company had had more rigid standards concerning the labeling of gaskets, for example, this accident might have been avoided.

OTHER ROOT CAUSES

Management Pressure

A root cause of some accidents is management pressure. Usually, there is pressure to increase production (or to get a plant started), leading to the possibility of safety or environmental performance being compromised. Although it is easy to criticize this attitude, it is more realistic to recognize that there will always be a balance between production and safety. After all, the safest plant is one that is not running at all. Just making the decision to operate is increasing risk. Managers never choose to operate unsafely; what they do is to balance economic and safety pressures. For example, whenever a manager is asked to give permission for a safety system to be bypassed so that maintenance can work on it, the manager is facing this trade off (*see* Chapter 16 for a discussion of the topic of acceptable risk.)

Because management pressure is often the most severe following a turnaround, when the plant has been down for a considerable amount of time, the Prestartup Safety Review element (Chapter 9) is important. It aims to prevent the temptation to cut corners, thereby jeopardizing safety.

Design Error

Design error is also something that often contributes to an incident but that is not discussed in the PSM program. The implicit assumption in most PSM work, particularly PHA's, is that the plant is built according to the correct standards, and that those standards are sound. Such an assumption may not be valid. For example, on many older plants there have been gradual increases in operating rates and conditions over a period of years. Yet an engineering study as to the capability of the relief and flare systems has not been carried out.

Related to the issue of design error is the problem that can occur with revamping and upgrading engineered systems. Such activities mean that old equipment may have to be validated against new codes; it cannot be "grandfathered." Hence, a minor upgrade or change in conditions can lead to major engineering changes throughout the system.

Willful Damage

Some incidents are created deliberately. The investigation of these is outside the scope of engineers and other technical people who work on the plant. The services of trained criminal investigators are needed, particularly if someone has been hurt.

PUBLIC INFORMATION REGARDING INCIDENTS

To this point, Incident Investigation has been discussed as being purely an internal matter. However, there is a good deal of information concerning incidents that is publicly available, and that can be of value to all facilities and plants. This is the message of Trevor Kletz' book, Lessons From Disaster, in which he argues that the public dissemination of accident and incident histories will help all companies improve their safety. For legal reasons, it may be necessary to change some of the information to protect individuals and to keep sensitive industrial information secret, but changing the information in this way is not likely to change the value of the lesson learned.

CONCLUSIONS

Incident Investigation — particularly when applied to near misses — is an invaluable way of improving safety because it is fundamentally non-theoretical. It provides insights as to what is really going wrong on a facility. The most difficult part of this element is simply getting people to report in incidents that may make them look bad. It is vital that management create a culture in which the reporting of incidents is a cause for praise, not blame.

Chapter 14

EMERGENCY PLANNING & RESPONSE

INTRODUCTION

 The PSM standard requires that companies prepare an emergency plan so as to minimize injuries and damage in the event of an accident. The importance and value of this element of the standard was demonstrated in the accident that occurred at a facility in Pennsylvania on October 16, 1995. This was a very serious accident, in which five people died. Yet, because a full-scale fire drill had been carried out at the site the day before the accident, one of the investigators was quoted in the Chemical Process Safety Report as saying: "This, along with the on-going emergency planning, are credited with preventing the incident from being far worse."

The process safety regulation to do with emergency planning and response says very little. Instead it refers companies to other OSHA standards which already cover this topic in detail, particularly §1910.38(a) (Employee Emergency Plans and Fire Prevention Plans) and §1910.165 (Alarm Systems.)

In most areas of process safety, there is very little difference between the OSHA PSM standard and the EPA RMP rule. However emergency response is an exception to this — there are significant differences between the two, largely due to the fact that the EPA is concerned with the impact of a release on members of the general public, whereas OSHA is more interested in worker safety.

THE OSHA REGULATION

The employer shall establish and implement an emergency action plan for the entire plant in accordance with the provisions of 29 CFR 1910.38(a). In addition, the emergency action plan shall include procedures for handling small releases. Employers covered under this standard may also be subject to the hazardous waste and emergency response provisions contained in 29 CFR 1910.120(a), (p) and (q).

OSHA GUIDANCE

Each employer must address what actions employees are to take when there is an unwanted release of highly hazardous chemicals. Emergency preparedness or the employer's tertiary (third) lines of defense are those that will be relied on along with the secondary lines of defense when the primary lines of defense which are used to prevent an unwanted release fail to stop the release. Employers will need to decide if they want employees to handle and stop small or minor incidental releases. Whether they wish to mobilize the available resources at the plant and have them brought to bear on a more significant release. Or whether employers want their employees to evacuate the danger area and promptly escape to a preplanned safe zone area, and allow the local community emergency response organizations to handle the release. Or whether the employer wants to use some combination of these actions. Employers will need to select how many different emergency preparedness or tertiary lines of defense they plan to have and then develop the necessary plans and procedures, and appropriately train employees in their emergency duties and responsibilities and then implement these lines of defense

Employers at a minimum must have an emergency action plan which will facilitate the prompt evacuation of employees when an unwanted release of highly hazardous chemical. This means that the employer will have a plan that will be activated by an alarm system to alert employees when to evacuate and, that employees who are physically impaired, will have the necessary support and assistance to get them to the safe zone as well. The intent of these requirements is to alert and move employees to a safe zone quickly. Delaying alarms or confusing alarms are to be avoided. The use of process control centers or similar process buildings in the process area as safe areas is discouraged. Recent catastrophes have shown that a large life loss has occurred in these structures because of where they have been sited and because they are not necessarily designed to withstand over-pressures from shockwaves resulting from explosions in the process area.

Unwanted incidental releases of highly hazardous chemicals in the process area must be addressed by the employer as to what actions employees are to take. If the employer wants employees to evacuate the area, then the emergency action plan will be activated. For outdoor processes where wind direction is important for selecting the safe route to a refuge area, the employer should place a wind direction indicator such as a wind sock or pennant at the highest point that can be seen throughout the process area. Employees can move in the direction of cross wind to upwind to gain safe access to the refuge area by knowing the wind direction.

If the employer wants specific employees in the release area to control or stop the minor emergency or incidental release, these actions must be planned for in advance and procedures developed and implemented. Preplanning for handling incidental releases for minor emergencies in the process area needs to be done, appropriate equipment for the hazards must be provided, and training conducted for those employees who will perform the emergency work before they respond to handle an actual release. The employer's training program, including the Hazard Communication standard training is to address the training needs for employees who are expected to handle incidental or minor releases.

Preplanning for releases that are more serious than incidental releases is another important line of defense to be used by the employer. When a serious release of a highly hazardous chemical occurs, the employer through preplanning will have determined in advance what actions employees are to take. The evacuation of the immediate release area and other areas as necessary would be accomplished under the emergency action plan. If the employer wishes to use plant personnel such as a fire brigade, spill control team, a hazardous materials team, or use employees to render aid to those in the immediate release area and control or mitigate the incident, these actions are covered by §1910.120, the Hazardous Waste Operations and Emergency Response (HAZWOPER) standard. If outside assistance is necessary, such as through mutual aid agreements between employers or local government emergency response organizations, these emergency responders are also covered by HAZWOPER. The safety and health protections required for emergency responders are the responsibility of their employers and of the on-scene incident commander

Responders may be working under very hazardous conditions and therefore the objective is to have them competently led by an on-scene incident commander and the commander's staff, properly equipped to do their assigned work safely, and fully trained to carry out their duties safely before they respond to an emergency. Drills, training exercises, or simulations with the local community emergency response planners and responder organizations is one means to obtain better preparedness. This close cooperation and coordination between plant and local community emergency preparedness managers will also aid the employer in complying with the Environmental Protection Agency's Risk Management Plan criteria.

One effective way for medium to large facilities to enhance coordination and communication during emergencies for on plant operations and with local community organizations is for employers to establish and equip an emergency control center. The emergency control center would be sited in a safe zone area so that it could be occupied throughout the duration of an emergency. The center would serve as the major communication link between the on-scene incident commander and plant or corporate management as well as with the local community officials. The communication equipment in the emergency control center should include a network to receive and transmit information by telephone, radio or other means. It is important to have a backup communication network in case of power failure or one communication means fails. The center should also be equipped with the plant layout and community maps, utility drawings including fire water, emergency lighting, appropriate reference materials such as a government agency notification list, company personnel phone list, SARA Title III reports and material safety data sheets, emergency plans and procedures manual, a listing with the location of emergency response equipment, mutual aid information, and access to meteorological or weather condition data and any dispersion modeling data.

ANALYSIS OF THE REGULATION

Although the Regulation itself is limited to references to other standards, the OSHA Guidance does provide some useful information for setting up an Emergency Plan. Some of the key points are as follows:

1. Employers need to decide what type of release their own employees can reasonably be expected to handle. For small releases, the best action to take is often to find the isolation valves on either side of the release, and to close them. For a major incident, operators need to close the unit EIV's (*see* page 96.) There is a balance here; on the one hand, having a person move into a situation to close the valves puts him at risk. On the other hand, if he does not take prompt action, the small release may become bigger, and could lead to many more people being affected. Some companies implement a rule that a person will always move away from a situation, not toward it, unless he is part of the emergency response team and is dressed in the appropriate protective clothing.

2. For each type of release and fire, employers must ensure that emergency evacuation routes are in place.

3. There must be an alarm system. OSHA §1910.165 provides information on how to set up such a system.

4. The Guidance discourages the use of control rooms as safe areas on the assumption that such buildings are not blast proof. If they are, then they would be an excellent place to assemble, depending on their proximity to the accident. If the danger is from the release of a toxic gas, it is important to make sure that the air conditioning for these buildings can be isolated so that the gas is not pulled into the control room. Some companies provide air packs inside the control room.

5. Guidance on the emergency release of hazardous chemicals is provided in 29 CFR 1910.120 (the HAZWOPER) standard.

LEVELS OF EMERGENCY

There are usually three levels in of emergency. The first level occurs when an employee notices a small incident, such as a leak or fire which he or she is able to bring under control with no assistance. The second level occurs when the employee recognizes that the situation is too large for one person to handle safely. If he were to attempt to do so, he could be hurt, thus adding to the scope of the problem. At this point, the plant emergency response team will take over. The personnel on this team will be specially trained in the handling of emergencies, and they will be issued with the proper clothing and equipment. If the situation becomes too large for the Emergency Response Team, then they can call for help from outside organizations, including the local fire department and/or other facilities in the area.

THE EMERGENCY ACTION PLAN

Each facility must develop an Emergency Action Plan. The elements of such a plan are described below.

ELEMENT 1 —ACCIDENT PREDICTION

The element in the plan is to predict the type and scope of accidents that can occur, usually with regard to fires, explosions and the release of toxic chemicals. A problem associated with the development of Emergency Plans is that there is such a wide range of possible accidents that can be considered. Factors to be included are the location of a release, its magnitude, wind direction and the number of people who may be in the area at the time of the release.

ELEMENT 2 —HAZARD MODELING

It can be very useful to model some of the scenarios, particularly the release of hazardous chemicals so that, if the accident actually does occur, the emergency responders will have some idea as to the size of the incident with which they may be expected to cope. Some companies even have on-line models which are constantly available. Then, if there is a release of that chemical, the response team can provide the modelers with current information so that a real-time prediction as to the magnitude of the incident can be developed. This has even reached the point where some fire departments carry computers on board their trucks so that they can model the release as they are driving toward it. This is particularly useful if they are expected to evacuate and/or treat members of the public.

ELEMENT 3 —PLANNING

Once the models and release scenarios have been prepared, the emergency response team should put together a plan such as that shown in Table 14-1. As with all other elements of PSM, this plan must be written down.

Table 14-1

Elements Of An Emergency Response Plan

I.	Identification of reasonably foreseeable emergency situations that could have serious adverse health impacts on the health and safety of employees or the public, or the environment.
II.	Instructions for managers, operators, other employees and contract workers, including the following:

A. Emergency organization
B. Alarm systems
C. Internal communications network
D. External notification requirements (media, community agencies and neighborhood contacts)
E. Emergency equipment and facilities
F. Specific steps for minimizing hard to human health and the environment and for returning to normal operations.

III.	Provisions for employee training and drills.
IV.	Provisions for notifying senior and/or corporate management.

The Emergency Action Plan should contain a description of the procedures to be followed in the event of an accident. It is particularly important to identify any chemicals that require special treatment. For example, the use of water on some chemicals may cause them to ignite. In these cases, they must be controlled with other chemical agents.

ELEMENT 4 — ORGANIZATION AND PERSONNEL

During the course of the emergency, a special reporting structure may be set up in which the Incident Commander is in complete charge of the facility; everyone else on the unit, including the normal management, report to the Commander. The Incident Commander will be in charge of the Emergency Response Team (ERT) which is made up of specially trained personnel from each shift.

In large industrial centers, such as the Texas Gulf Coast, the various plants coordinate their emergency response efforts in a mutual support system. So, if a facility has a fire and needs additional fire-fighting trucks, they will be supplied by neighboring units. If the incident is bad enough, there will be a ripple effect of emergency equipment moving toward the affected site for dozens of miles.

In any major event, the local fire department and ambulance services will probably be involved. They should have the opportunity of training with the plant emergency response team.

The persons responsible for running the unit should always know how many people are on the site at any one time. For larger facilities, they should also know roughly where those people are within the facility. If there is a major accident, an accurate head count will determine if anyone needs rescuing.

The Emergency Plan must identify all the equipment that the emergency responders have the equipment that they need to do their work properly. It should also indicate where the equipment is located.

ELEMENT 5 — INVESTIGATION

If the incident is serious, an investigation as to what caused it will start as soon as everyone is out of danger. It is particularly important to find out what happened if there are reasons to believe that it could happen again, maybe at another site.

ELEMENT 6 — RECOVERY OPERATIONS

As soon as the site is secure, and there is no danger to anyone, recovery of equipment and chemicals can start. At this time, the plant may contain many unexpected hazards, such as the danger of being struck by falling equipment that has had its foundations damaged by fire. Also, some equipment may be contaminated with hazardous chemicals, and may need to be specially treated before it can be returned to service, or before the operators or maintenance personnel can use it.

ELEMENT 7 — BRIEFING THE PRESS, PUBLIC, OSHA

It is important that the press and the public be informed of what is going on at the site, particularly if anyone is in any danger. Telephone lines and other links for public communication must be available, and that they do not become jammed with unnecessary calls.

CONCLUSIONS

Like the topic of Hot Work, Emergency Response & Planning is not usually a critical part of a new process safety program because the systems are already in place, and are tested on a regular basis. However, it is important to make sure that the response systems are exercised on a frequent basis, and to make sure that they conform to the appropriate OSHA standards.

Chapter 15

COMPLIANCE AUDITS

INTRODUCTION

 The purpose of an audit is to compare actual performance with written, pre-defined standards. The standards may be external, such as a government regulation, or they may be internal. Wherever they came from, and whatever their purpose, the standards must be written down. No matter how well managed a facility may be, there will always be areas of deficiency, where the facility is not living up to its own standards. Having compared actual operation with the standards, the auditor issues a report on what he or she has found. Then the client management uses the findings in the report — possibly with the assistance of the auditor — to improve management systems.

Audits must be performed on a regular basis (with regard to process safety, the OSHA standard specifies that they must be done at least every three years.) However, like all elements of process safety, audits should be an on-going activity performed by everyone all the time. Everyone should be constantly looking for and reporting on problems, including deficiencies in his or her own actions. Whenever a person notices that the plant operation is deviating from design or operating intent, regardless of whose fault it is, that person should note the problem and take appropriate action. An example of this was provided on page 65 regarding the operator who decided not to open a valve until the consequences of this action had been analyzed more carefully.

The most serious audits are those conducted by a regulatory agency, or by attorneys for the plaintiffs following an accident. However, most audits are voluntary; a company decides that it needs to check up on itself, even though there is no reason to believe that a serious problem exists. Some of these voluntary audits are carried out as if they actually are being conducted by regulatory bodies, i.e. the auditors behave as if they are regulators who are investigating an accident. Not only does this help identify safety problems, it may also reduce a company's exposure to penalties should there actually be an accident because it demonstrates the company's commitment to doing a good job. It also familiarizes the plant personnel with what a real audit might be like should there ever actually be an accident.

A formal audit does not have to be hostile in tone. Nevertheless, there is no question that its objective is to identify and report on deficiencies. No matter how discretely the final report may be worded, and no matter how courteous the auditor may be while conducting his or her investigation, it must never be forgotten that the purpose of a formal audit is to find deficiencies. The auditor is not responsible for suggesting ways of improving performance — he or she is simply checking performance against a standard.

If an auditor is asked to help with providing solutions, then the audit has become a review, sometimes referred to as an assessment (Ozog et al.). Reviews are less formal than audits. The reviewer typically works as part of a team with the facility personnel. Even though looking for problems, he or she is not out to find fault. In particular, there is no threat of penalties or retributive action based on his findings. Moreover, the findings of a review can be considered against the reality of budgets and schedules. So, if the reviewer identifies ten deficiencies, they can be prioritized based on a knowledge of budgets, personnel availability and turnaround schedules. (An formal audit would not do this; it would simply report on the ten deficiencies.) Finally, a reviewer will often work with the facility being audited on developing solutions to the problems that have been identified.

The terms Verification and Validation can also be used to distinguish between auditing and reviewing. Verification is concerned with ensuring that a facility meets the letter of the law or regulation; Validation determines whether it is meeting the spirit of the same law.

It is useful to have a mix of both formal audits and reviews; there is a place for both. What is important is to make a clear distinction between them, and to understand which is being carried out at any one time. This is not always easy. As Kenney (304) points out, if a process plant auditor is to ask meaningful questions, he has to have extensive plant experience. (This is in contrast to say financial audits, which do not generally require that the auditor have in-depth knowledge of their clients' businesses.) The catch with using experienced people for process safety audits is that they are likely to be sympathetic with the persons that they are auditing. Hence, they will want to try to help, not to evaluate.

Furthermore, the process industries traditionally have not been audited as often as other industries such as food processing or nuclear power generation. Therefore the people being audited may find the process to be somewhat threatening. (This situation is gradually changing because audits are becoming a more routine part of the process industries culture, largely due to the large number of environmental regulations that have been promulgated, and the spread of quality programs such as such as ISO 9000 and 14000.)

An audit of a PSM program faces a special problem: the lack of pre-defined, objective standards. As discussed in Chapter 1, process safety is performance-based; the only true measure of success is lack of accidents. The specific goals that derive from this larger goal have to be derived by each facility to meet their particular needs and circumstances. Therefore, the audit of a PSM program must start by establishing that specific goals have been developed for each element of the standard.

A concern sometimes expressed about audits is that they create "smoking guns", i.e. they identify problems which must then be addressed in a timely manner. Failure to do so means that the company is likely to found guilty of a willful violation should there then be an accident in which that deficiency was implicated, or if there is a follow-up audit. However, implicit in the decision to conduct an audit is the commitment to address any problems that are uncovered. This is why, in some respects, the most important person involved in the audit process is the manager in the host company who decides that an audit is needed. After all, an audit rarely results in universal praise, no matter how good a company's record may be; the auditor will certainly identify problems. It takes force of will to state that an audit is needed, knowing that the report will identify deficiencies, and that it will generate additional work for everybody.

The people being audited should always keep in mind the importance of the impressions that they make. An auditor will be impressed not only by a high quality PSM program, but also by one that looks good, i.e. one where the documentation is neat and tidy, and it is easy for the auditor to find his or her way around the system. (The preparation of a "good-looking" PSM program is discussed in Chapter 17.)

Finally, it is important for the auditors to stress their positive findings. If they find that something is being done exceptionally well, this should be pointed out, and the appropriate personnel recognized. This is the only time when it is appropriate to name names. Not only is the highlighting of positive findings good for morale, the identification of strengths may suggest ways of improving those parts of the program that are not so advanced.

THE OSHA REGULATION

The following is the PSM regulation (29 CFR 1910.119) with respect to Audits.

---

(1) Employers shall certify that they have evaluated compliance with the provisions of this section at least every three years to verify that the procedures and practices developed under the standard are adequate and are being followed.

(2) The compliance audit shall be conducted by at least one person knowledgeable in the process.

(3) A report of the findings of the audit shall be developed.

(4) The employer shall promptly determine and document an appropriate response to each of the findings of the compliance audit, and document that deficiencies have been corrected.

(5) Employers shall retain the two (2) most recent compliance audit reports.

OSHA GUIDANCE

Employers need to select a trained individual or assemble a trained team of people to audit the process safety management system and program. A small process or plant may need only one knowledgeable person to conduct an audit. The audit is to include an evaluation of the design and effectiveness of the process safety management system and a field inspection of the safety and health conditions and practices to verify that the employer's systems are effectively implemented. The audit should be conducted or lead by a person knowledgeable in audit techniques and who is impartial towards the facility or area being audited. The essential elements of an audit program include planning, staffing, conducting the audit, evaluation and corrective action, follow-up and documentation.

Planning in advance is essential to the success of the auditing process. Each employer needs to establish the format, staffing, scheduling and verification methods prior to conducting the audit. The format should be designed to provide the lead auditor with a procedure or checklist which details the requirements of each section of the standard. The names of the audit team members should be listed as part of the format as well. The checklist, if properly designed, could serve as the verification sheet which provides the auditor with the necessary information to expedite the review and assure that no requirements of the standard are omitted. This verification sheet format could also identify those elements that will require evaluation or a response to correct deficiencies. This sheet could also be used for developing the follow-up and documentation requirements.

The selection of effective audit team members is critical to the success of the program. Team members should be chosen for their experience, knowledge, and training and should be familiar with the processes and with auditing techniques, practices and procedures. The size of the team will vary depending on the size and complexity of the process under consideration. For a large, complex, highly instrumented plant, it may be desirable to have team members with expertise in process engineering and design, process chemistry, instrumentation and computer controls, electrical hazards and classifications, safety and health disciplines, maintenance, emergency preparedness, warehousing or shipping, and process safety auditing. The team may use part-time members to provide for the depth of expertise required as well as for what is actually done or followed, compared to what is written.

An effective audit includes a review of the relevant documentation and process safety information, inspection of the physical facilities, and interviews with all levels of plant personnel. Utilizing the audit procedure and checklist developed in the preplanning stage, the audit team can systematically analyze compliance with the provisions of the standard and any other corporate policies that are relevant. For example, the audit team will review all aspects of the training program as part of the overall audit. The team will review the written training program for adequacy of content, frequency of training, effectiveness of training in terms of its goals and objectives as well as to how it fits into meeting the standard's requirements, documentation, etc. Through interviews, the team can determine the employee's knowledge and awareness of the safety procedures, duties, rules, emergency response assignments, etc. During the inspection, the team can observe actual practices such as safety and health policies, procedures, and work authorization practices. This approach enables the team to identify deficiencies and determine where corrective actions or improvements are necessary.

An audit is a technique used to gather sufficient facts and information, including statistical information, to verify compliance with standards. Auditors should select as part of their preplanning a sample size sufficient to give a degree of confidence that the audit reflects the level of compliance with the standard. The audit team, through this systematic analysis, should document areas which require corrective action as well as those areas where the process safety management system is effective and working in an effective manner. This provides a record of the audit procedures and findings, and serves as a baseline of operation data for future audits. It will assist future auditors in determining changes or trends from previous audits.

Corrective action is one of the most important parts of the audit. It includes not only addressing the identified deficiencies, but also planning, follow up, and documentation. The corrective action process normally begins with a management review of the audit findings. The purpose of this review is to determine what actions are appropriate, and to establish priorities, timetables, resource allocations and requirements and responsibilities. In some cases, corrective action may involve a simple change in procedure or minor maintenance effort to remedy the concern. Management of change procedures need to be used, as appropriate, even for what may seem to be a minor change. Many of the deficiencies can be acted on promptly, while some may require engineering studies or in depth review of actual procedures and practices. There may be instances where no action is necessary and this is a valid response to an audit finding. All actions taken, including an explanation where no action is taken on a finding, needs to be documented as to what was done and why.

It is important to assure that each deficiency identified is addressed, the corrective action to be taken noted, and the audit person or team responsible be properly documented by the employer. To control the corrective action process, the employer should consider the use of a tracking system. This tracking system might include periodic status reports shared with affected levels of management, specific reports such as completion of an engineering study, and a final implementation report to provide closure for audit findings that have been through management of change, if appropriate, and then shared with affected employees and management. This type of tracking system provides the employer with the status of the corrective action. It also provides the documentation required to verify that appropriate corrective actions were taken on deficiencies identified in the audit.

ANALYSIS OF THE REGULATION

(1) Certification

The OSHA standard requires that an audit be carried out at least every three years. Failure to do will likely lead to a willful citation from OSHA (Elveston.) Some elements of the PSM program may need to be audited more frequently depending on the degree of risk that they represent.

(2) Technical Qualifications

The standard requires that at least one of the investigators should be knowledgeable in the process. Usually, this expertise will be provided by a process or operations expert who works on the unit being audited. If the audit is large and/or complex, specialists can be called in to review specific areas, such as electrical or instrumentation systems.

The OSHA guidance does not require that the audit team include someone from outside the organization or facility being audited. Nevertheless, it is often a good idea to use such a person. He or she will not be affected by internal personality issues, and, if a major problem is identified, will be able to report it without fear of retribution. Also, an outsider will be able to bring his or her experience from other companies to provide a reference against which to evaluate the present facility. Another reason for bringing in an outsider is that it is unusual for an operating facility to have someone on the staff who is an expert in the audit process.

(3) Report

The end product of all audits is a report. The layout of a typical report is discussed later in this chapter.

(4) Response

The standard requires that any findings from the audit are properly addressed. It is important to make sure that, once a problem has been resolved, the resolution is properly documented.

(5) Retention Of Reports

The audit reports must be retained as specified. They become part of the facility's Process Safety Information.

THE BASICS OF AUDITING

THE BASIC QUESTIONS

The five basic questions associated with any audit are: Who? What? When? Where? and How? With regards to process safety, these can be interpreted as follows.

Who

The auditor has to determine who is responsible for developing and implementing each element of the standard.

What

The auditor has to determine what the requirements are for each element in the context of the facility being audited. Sometimes, this is straightforward. However, it often requires judgment as to what should be included, which means that the auditor must be experienced both in auditing and in process plant operations.

When

As already discussed, audits should be carried out all the time. All employees and contract workers should be constantly evaluating the safety of the facility at which they are working. For formal audits, the three year cycle require by OSHA is suitable for many companies and sites.

Where

The appropriate process safety actions may vary at different parts of a facility. For example, one section may contain a very hazardous chemical that is not found in other sections.

How

The auditor must ensure that the element is being implemented properly. This often requires good judgment. An important part of this question is to make sure that any findings — either from the audit or from other work, such as PHA's — is closed out in a proper manner.

THE AUDIT PROCESS

The audit process will generally proceed as follows.

1. Determine the standards that the facility is following (regulations, industry codes and internal.)
2. Gather information and cross-check it against other, independent sources.
3. Identify individual problems that are not systemic.
4. Identify problems that are indicative of deficiencies in management systems.
5. Derive conclusions as to what the management system problems may be.
6. Provide guidance for correction (optional.)

The first step is to identify all the rules and standards that the facility is following with regards to process safety. Some of these, such as OSHA and EPA regulations, will be self-evident. Others, such as the company's own culture and expectations may be harder to define.

Having determined the standards, information must be collected. This usually consists of collecting and reading documents, interviewing operators and maintenance personnel and conducting a plant tour. (During the tour, the auditor should inspect areas where which may normally be overlooked, such as the tops of large columns or behind the control panels.) Wherever possible, auditors should try to cross-check information with information from a different source. For example, the auditor can check an operating procedure by asking an operator to demonstrate how a particular task is carried out in the field. Or he can check the Management of Change system by ensuring that a field change that is currently taking place was properly analyzed, and that it was marked up on the P&ID's.

The second step is to identify individual problems that are not systemic. In the definition of audits provided at the start of this chapter, a key phrase was "management systems." Sometimes, accidents are caused by isolated issues. For example, it may be found that one particular individual has domestic problems that affect his safety performance. If this person is removed or reassigned, that particular problem will be removed. (However, even isolated issues such as this may be symptomatic of management problems. For example, it may indicate that the hiring or training programs need improvement, or that some type of company-sponsored, employee help program is needed.) If individual or unique problems are identified, they can usually be addressed promptly.

The next step is to identify those issues that indicate the existence of broader management problems. There are two items that help indicate this. The first is the occurrence of similar problems in different areas, or at different facilities. For example, if piping failure due to corrosion is widespread, it is likely that there is fundamental problem with at least part of the Mechanical Integrity program. Another signifier of management problems is inconsistency between elements. If the plant has a good operating procedures program, for example, but it is found that the training program is not as good, there may be a management issue to do with transferring knowledge from one group to another.

Having identified the existence of management problems, it is then necessary to carry out some type of root cause analysis to determine what it is. For this, the methods described in Chapter 13 (Incident Investigation) may be useful.

The final step in the audit process — provide guidance — is not always required. Strictly speaking it is not part of the formal audit as such; instead, the auditor is providing suggestions as to how corrective action can be taken based on his or her observation of how other plants operate. Regardless of the fact that this may not strictly speaking be an audit step, it is what many clients want — particularly if the auditor is from an outside company — because comparisons with other companies and facilities can be extremely valuable, and can stimulate creative and insightful thinking.

The auditor must always remember that any guidance that he or she offers is based on opinion. If the auditor finds that there is a procedure missing, for example, he will note that deficiency, and there can be no argument. However, if the auditor states that, in his opinion, the procedure is too short (or too long) he will have to defend that opinion, and indeed may have to change his mind if the client can demonstrate that what they have leads to a safe and efficient operation.

The problem of judgment also occurs when a company has more than one facility performing essentially the same operation, but their safety programs are different. The auditor may be called in to determine which of the two is better.

THE AUDIT STAFF

TYPES OF AUDITOR

An auditor must always be independent of the facility being reviewed and must always have complete freedom to observe and report on all aspects of the operation. The outside perspective not only ensures that there is no conflict of interest. It can also provide alternative points of view that can be helpful to the people working within the facility.

Outside auditors generally come from one of the following four groups.

1. A regulatory agency.
2. Attorneys representing plaintiffs following an accident.
3. An outside company that specializes in process safety work.
4. A person from within the client company, but based in another department.

Of these four, audits by an agency or by attorneys are the most challenging. If the audit is being carried out because there already has been an accident or because there have been complaints, an adversarial atmosphere exists from the start.

The use of outside companies to lead the audit offers three benefits to the facility being examined. First, the auditors will be objective. They are not familiar with the processes being audited, nor are they involved with the inter-personal relationships on the unit. For them, this project is only one of many. Therefore, if their findings are negative, they will be much less reluctant to say so than would a company employee. In other words, an outside auditor is not likely to be as affected by company politics.

The second benefit of using an outsider is that they will bring a fresh perspective based on their knowledge of how other companies and facilities operate. In particular, they may identify hazardous situations that employees at a facility have learned to live with but that would not be accepted as being safe elsewhere.

Finally, an audit conducted by an outsider will give the feel of a full regulatory audit. The fact that the person(s) involved are strangers, and that they have no particular ax to grind, will create a sense of what an audit could be like were there to be a bad accident.

Large companies sometimes appoint auditors from other departments or facilities within that company. These people should act as if they are completely independent of the facility that they are auditing. In practice, this can be difficult, even if the company is large enough to be able to afford a separate corporate group whose only function is to audit their own company's plants. Auditors from other departments generally face two problems that true outsiders do not have to worry about. The first is that the company as a whole may have some blind spots that are common to all of their facilities. Every company has its own culture, and it may be that internal auditors will not realize that such blind spots exist. The second problem that internal auditors may have to overcome is their own knowledge of company problems and issues. These can affect the auditor's objectivity. Some of this may not be explicit, and it may even be unconscious, but it is still a factor. For example, if the internal auditor knows that his company is in financial trouble, he may go easy with the audit, whereas a true outsider will focus on just the safety issues with which he or she is confronted.

TEAM MEMBERS

A short audit of a simple unit may be carried out by a single individual who is knowledgeable in the process. If the scope of the audit is larger, specialists will be needed to make sure that the right questions are being asked (rather like a PHA.) Specialties that may be included are:

- Process engineering
- Electrical engineering
- Instrument engineering
- Maintenance/Mechanical engineering

- Quality assurance
- Risk analysis

THE HOST COMPANY

It is often necessary to brief those people in the company that is being audited regarding how to work with the auditors. Items to be considered include:

- The purpose of the opening meetings.
- How to answer all questions fully and honestly, but being careful about volunteering information that does not pertain to the question being asked. There should also be a discussion as to what gifts (if any) are appropriate. Inspectors from some government agencies, for example, cannot accept even a cup of coffee (*see* page 22.)
- How to respond when deficiencies or problems are uncovered.
- Whether to allow the auditor independent access to people and records, or whether the auditor should always be accompanied by a staff counterpart, who will serve as a watch-dog.
- When to refuse to answer questions on grounds of not having sufficient authority or knowledge to make a trustworthy reply.
- Knowledge of which records are open for inspection.
- Whether to take matching information. For example, if the auditor takes pictures, his or her escort must know whether to take a picture of their own at the same time so that there is independent verification of the auditor's findings.

Immediately following the inspection phase of the audit, the facility that was audited may choose to issue a brief internal report that documents their assessment of the audit, and that records all the independent information that they collected.

PERSONAL ATTRIBUTES

Audits are stressful for those being audited. No matter how good a plant's performance may be, there is every chance that the auditor is going to find something that indicates a problem; and all problems and deficiencies are eventually traced back to particular individuals. Therefore, the personal attributes and conduct of the auditor is extremely important. He or she must immediately establish their credentials, which usually means demonstrating knowledge and skill in the three following areas:

a) Audit protocols
b) Plant operations
c) Pertinent regulations and standards

Throughout the audit, the auditor must be friendly and empathetic without losing sight of the fact that he or she is an outsider, and that they must stand apart from the personalities at the client site.

It is suggested that the auditor does not make any negative comments during the audit process, but he or she should always express appreciation for the effort that the client is making to provide information and assistance.

Finally, the auditor must be careful not to personalize his or her work. The identification of a deficiency does symbolize "success", nor is it a personal triumph. The auditor's job is to report facts as dispassionately as possible.

THE ROLE OF THE AUDITOR'S OPINION

An auditor who has considerable experience of process operations will inevitably have personal opinions about the operation that he or she is auditing (this is the Validation process.) For example, if the operating procedures on the facility being audited are much shorter than the operating procedures on other similar facilities, he or she will investigate this element with increased scrutiny. The shortness of the procedures may indicate that facility has a deficiency in this area. Therefore, although an auditor is not allowed to explicitly state an opinion (all comments must be an evaluation of facts against a standard), he or she has can use experience to help guide and direct the course of the audit.

IMPLEMENTATION

A typical audit can be divided into the following steps.

1. Plan The Audit
2. Conduct The Audit
3. Report
4. Evaluate The Findings/Take Corrective Actions

PLAN THE AUDIT

GOALS

Before the audit starts, the facility management must decide what it intends to gain from the audit in addition to simply verifying that they are in conformance with a regulatory standard. For example, the auditors can be asked to measure compliance with non-regulatory standards such as ISO 9000. Also, the audit team needs to know whether the present audit is meant to evaluate compliance with the recommendations and findings of the previous audits (where they exist.)

Topics that can be raised while setting goals include the following:

- The business objectives of the plant
- Other objectives, such as safety or environmental goals, or other company policies
- Management philosophy and organization
- PSM objectives
- Regulations and standards that are pertinent to the audit

PHYSICAL SCOPE

Generally, the physical scope of the audit is well understood. Usually, it is confined to one or more facilities, each of which has well-defined boundaries. However, there can be complications, such as whether areas and functions such as the following are to be included:

- Utility systems
- Vendor Equipment
- Vehicle movement
- Instrument and control loops
- Human factors

DESIGN OF THE AUDIT FORMS

The audit planning process includes the design of the audit forms. These forms will usually contain four major sections:

1. A question that refers back to the appropriate standard.
2. A check box that provides space for the answer to the above question.
3. Space for recording where the information came from.
4. Space for discussion and comments.

Usually, each question has one of four possible answers:

1. Satisfactory — Meets Requirements
2. Unsatisfactory — Does Not Meet Requirements
3. Not Audited
4. Not Applicable

(The response of "Partially Satisfactory" is not allowed — it is the same as Unsatisfactory.)

OSHA AUDIT GUIDELINES

OSHA has prepared a set of publicly available audit guidelines and questionnaires for use by their own Compliance Safety and Health Officers (CSHO's.) Generally, these audit guidelines mimic the wording of the standard very closely. Frequently, the wording of the standard is simply reversed in order to create the audit question. These guidelines (OSHA Instruction CPL 2-2.45A) were issued on September 13, 1994. Table 15-1 provides an example of one of these guidelines (in this case, for the Employee Participation element.)

Table 15-1

Sample OSHA Audit Form

1910.119 (c) EMPLOYEE PARTICIPATION

I. PROGRAM SUMMARY

The intent of this paragraph is to require employers to involve employees at an elemental level of the PSM program. Minimum requirements for an Employee Participation Program for PSM must include a written plan of action for implementing employee consultation on the development of process hazard analyses and other elements of process hazard management contained within 1910.119. The employer must also provide ready access to all the information required to be developed under the standard.

II. QUALITY CRITERIA REFERENCES

A. 1910.119(c): Employee Participation

III. VERIFICATION OF PROGRAM ELEMENTS	CRITERIA REFERENCE	MET Y/N
A. Records Review 1. Does a written program exist regarding employee participation? Field Note Reference(s):	.119(c)(1)	

III. VERIFICATION OF PROGRAM ELEMENTS	CRITERIA REFERENCE	MET Y/N
2. Does the written program include consultation with employees and their representatives on the conduct and development of other elements in the PSM standard? Field Note Reference(s):	.119(c)(2)	
3. Does the written program provide employees (including contractor employees) and their representatives access to process hazard analyses and all other information developed as required by the PSM standard? Field Note Reference(s):	.119(c)(3)	
B. On-site Conditions Not applicable		
C. Interviews 1. Based on interviews with a representative number of employees and their representatives, have they been consulted on the conduct and development of the process hazard analyses? Field Note Reference(s):	.119(c)(2)	

III. VERIFICATION OF PROGRAM ELEMENTS	CRITERIA REFERENCE	MET Y/N
2. Based on interviews with a representative number of employees and their representatives, have they been consulted on the development of other elements of the Process Safety Management program? Field Note Reference(s):	.119(c)(2)	
3. Based on interviews with a representative number of employees (including contractor employees) and their representatives, have they been informed of their rights of access and provided access to process hazard analyses and to all other information required to be developed by the PSM standard? (Ask about unreasonable delays in access to information and whether time is given during the working hours to access information required by the PSM standard.) Field Note Reference(s):	.119(c)(3)	

PRE-AUDIT ACTIVITIES

Before the audit starts, the host company will probably spend some time preparing for their visitors (unless the audit is unannounced.) As has already been discussed, this can be a useful time for making corrections to items that obviously require attention. In other words, the very act of having an audit can lead to improvements in safety. On the downside, these activities can lead to the development of a skeptical attitude on the part of those being audited. They will concentrate on "passing the audit" rather than on running a safe operation.

CONDUCT THE AUDIT

KICK-OFF MEETING

The audit will start with a kick-off meeting. The audit team and representatives from the client company will meet to discuss the way in which the audit is to be carried out. Basically they will discuss four items:

1. The objectives of the audit — why it is being performed.

2. The mechanics of the audit: including issues such as schedules, budgets and facilities to be covered.

3. The scope of the audit: including issues such as regulations to be covered and how deeply the research should go.

4. The depth of the audit questions on technical issues.

The kick-off meeting is not a one-time affair. Every time a new person from the facility is brought into in the audit, he or she should have their own kick-off meeting that provides them with an explanation as to what is going on, and what is expected of them. In particular, it must be made clear to those people who have not been through an audit before that the aim is not to find fault with individuals, but to identify weaknesses in the management systems.

During the kick-off meeting, background documentation, including items such as those listed below, should be available.

1. Permits.
2. Incident reports.
3. Logbooks and DCS records.
4. MSDS and other safety information.
5. An organization chart
6. A plot plan, including information on population concentrations such as schools, hospitals and prisons.

7. Process description.
8. Previous audits and inspections (including the action items that they generated.)
9. Safe work permit procedures.

THE AUDIT

The audit is guided by pre-designed protocols and checklists and is carried out using a combination of interviews and documentation reviews. Typically, the one verifies the other. For example, the auditor may ask an operator how he performs a certain task (the interview.) Having been told, he will then ask for a written procedure that confirms this (the written step.) This can work the other way around just as well. The auditor can read an MSDS, and then ask the operator to verbally confirm that he knows the major safety issues to do with the chemical in question.

In all but the most rigorous audits (i.e. those following a major accident), the auditor will have to be selective regarding the documents that he reads, and the people chosen for interviews. There simply is not enough time to read and discuss everything. This is where experience is invaluable. An experienced auditor will sense where the soft spots are and, based on the information received to that point, will use his or her experience to decide on the follow-up questions.

It is also important to obtain information from as many independent sources as possible. For example, the training records should be reviewed not only with the training supervisor, but also with one or more of the operators.

INTERVIEWS

The interview is at the heart of most audits. The auditor systematically works through a series of questions with one or more of the people who work at the site. Typically, the process follows these steps:

1. The auditor asks the interviewee to describe his work, its scope and the areas is covers. This gives the auditor a sense of the boundaries for this particular interview.

2. The auditor then works through a questionnaire that corresponds to the PSM standard. So, the OSHA standard with respect to emergency response, for example, requires that "the emergency action plan shall include procedures for handling small releases." The auditor will ask to see the emergency plan, and will ensure that it contains provision for handling small releases.

3. The auditor will then pursue the question in more detail, often using the other elements of the standard for guidance. So, using the above example of emergency response to small releases, the auditor will ask the interviewee for copies of the procedures to handle this scenario, evidence that the operators have been trained in it, documentation that contract workers know what to do in the event of a small release, and so on.

4. The auditor will then move on to the next question. He or she is always watching out for systemic problems. If one particular procedure is not in place, that may be an oversight. If, however, it is found that there is systematic lack of emergency procedures, this indicates a more fundamental management problem. The auditor can also use the first answers as guidance for where to focus later on in the audit.

The breadth of questioning is a very important consideration. Unless there has been a serious accident, when all factors will be considered, it is likely that most audits will work on a spot-check basis. Certain procedures and actions will be selected at random, and then evaluated by the investigator. Two problems that this creates are:

1. How are procedures and activities selected so as to be representative of the overall picture?

2. Have sufficient procedures and activities been selected, and have they been selected from different operating areas, which may have different standards from one another?

During the interview the auditor must take notes, otherwise he will not be able to verify his findings. The note-taking should be as discrete as possible, so that the interviewee is not distracted or worried. The use of tape recorders is probably not a good idea because most people will find them to be threatening.

One problem that sometimes occurs with regards to interviews is that the interviewee's manager or supervisor may wish to be present during the conversation. It is up to the auditor to decide if this is acceptable. The presence of a manager may inhibit the interviewee, and prevent him from describing the situation as it is, rather than as it should be. Moreover, the manager may attempt to direct the audit in ways that make him look favorable. This is particularly the case with issues to do with communication and training because of a possible gap between theory and reality. A manager may point out that written policies and programs are available for these topics, but he may be reluctant to have an auditor speak to an operator, only to find that the operator is completely unaware of these written policies.

ON-SITE INSPECTION

The auditor should verify information by field inspection. For example, an operator may have a procedure for carrying out a task, but, when demonstrating what he actually does when working with the plant equipment, may demonstrate that his actions are not in conformance with the procedure.

THE ORDER OF THE ELEMENTS

Generally, the order in which the elements are audited is not very important. Also, logistics often force a particular schedule to be implemented. For example, if the plant maintenance engineer is only available on a particular day, then that will be the time to review mechanical integrity. However, the order in which items are discussed may be directed by the results found to that point. For example, if a problem to do with the emergency procedures is found, the auditor may suspect that the same problem will occur with the operating procedures. Hence, he may choose to review both of these items at the same time.

Leaving these considerations aside, the auditor should start with those items that are likely to have general application throughout the entire facility. That way, if he finds a problem, his subsequent investigation will serve to confirm or deny that first impression. For most plants, therefore, the following three items provide a starting point:

- Employee Participation because of its central importance to process safety.
- P&ID's because they provide the informational base.
- Management of Change because of its criticality in preventing uncontrolled change.

OSHA RECORDS

Companies are required to keep records of occupational-related illnesses and injuries. These are normally recorded in the OSHA 200 logs. These records provide an excellent review of overall safety performance, and can provide the basis for an Incident Investigation program.

EXIT MEETING

At the conclusion of the audit, the audit team should have an exit meeting with the host company. This will provide an opportunity for the auditors to provide an overview of their results and to highlight any serious issues that may have been identified and that require immediate attention. The response of the facility personnel to these preliminary findings can be incorporated into the final audit report.

The auditor should try not to say too much at this stage because he may find that his conclusions change as he is writing the report. Hence, any preliminary statements may have to be modified or even retracted. This makes the auditor look unprofessional, and it may cause the client to initiate inappropriate corrective actions.

THE AUDIT REPORT

Once the audit is finished, the auditor must write a report on what he has found. Typically, there are four steps in this process:

1. Debrief the client.
2. Prepare a draft report.
3. Review the report with the client.
4. Issue the final report.

THE DRAFT REPORT

The auditor should prepare the draft report as quickly as possible. No matter how effective the auditor's note-taking may have been, there will always be some observations and facts that were not properly written down. These should be captured before the auditor forgets them. It also gives the client an early opportunity to correct any factual errors in the report.

There is a well-known proverb in the consulting business that "There is no such thing as a draft report." Draft reports may contain errors of fact, and will probably look relatively disorganized and unprofessional. The client will probably fixate on these problems. Then, no matter how high the quality of the final report, an indelible image of poor quality will remain.

It is important that the auditor has someone within his or her own company review the report before the client sees it. The purpose of this review is to catch any obvious, and potentially embarrassing, errors. Items to watch for include:

*   Inconsistency between reported facts
*   Inappropriate language
*   Opinions masquerading as facts
*   Completeness, particularly with respect to the standard (such as OSHA's PSM) being used

The draft report should be issued to the client as soon as possible. Where feasible, the report should be presented in a meeting, particularly if it contains difficult or sensitive findings. The auditor should expect to be questioned and challenged as to his conclusions, and it is best if he can do this face to face.

Sometimes a client will wish to add information concerning action that was taken following the audit. This is often the case if the auditor had identified a serious problem that was corrected right away. This type of information should not be included in the audit report itself (which is a report of the plant at a particular time.) However, it is not unreasonable to have a paragraph in the report discussing follow-up actions that have been carried out, and that are planned.

THE FINAL REPORT

Table 15-2 provides an outline of the Table of Contents for an audit report. It is similar in some respects to the Chemical Safety Audit form prepared in 1988 (Ross.) The details will vary according to the facility and why the audit was being conducted.

Table 15-2

Table Of Contents For Representative Audit Report

Letter of Certification

I. **Introduction**
  A.   Reason For The Audit
  B.   Pertinent Regulations
       1.   29 CFR 1910.119
       2.   40 CFR 68
       3.   etc.
  C.   Permits Examined
  D.   Pertinent Standards
       1.   API Recommended Practice 750
       2.   ISO 9001
       3.   etc.
  E.   Goals And Objectives
  F.   Audit Methodology
  G.   Audit Scope — Units Covered
  H.   Audit Team
  I.   Documents And Records
  J.   Circulation Of The Report

II. **Facility Information**
  A.   Site Description
  B.   Location
       1.   Address
       2.   Local Community
       3.   Local Emergency Planning Committee
       4.   Contact Information
  C.   Facility Ownership
  D.   Facility Management
  E.   Facility History
       1.   Operations
       2.   Safety
       3.   Environmental
       4.   Emergency Response And Planning Program
           a)   Escape Corridors
           b)   Community Communication Program

        c)   Emergency Response Capability —
                 On-Site
        d)   Emergency Response Capability —
                 Off-Site
        e)   Emergency Chain Of Command
        f)   Emergency Equipment
        g)   Training For Emergency Response

F.   Safety And Loss Prevention Program
    1.   Corporate Policy
    2.   Site Policy
    3.   Industry Guidance And Standards

III.   Hazardous Chemicals On Site

A.   Chemical 1
    1.   Location
    2.   Properties
    3.   Amount Stored

B.   Chemical 2
    1.   Location
    2.   Properties
    3.   Amount Stored

C.   etc.

IV.   Process Description

A.   Process Overview

B.   Major Process Areas
    1.   Area 100
    2.   Area 200
    3.   etc.

C.   Location Of Hazardous Chemicals
    1.   Chemical 1
    2.   Chemical 2
    3.   etc.

D.   Major Process Equipment
    1.   Area 100
    2.   Area 200
    3.   etc.

V.   Process Safety
    A.   Elements Covered
                1.   Employee Participation
                2.   Process Safety Information
                3.   Process Hazards Analysis
                4.   Operating Procedures
                5.   Training
                6.   Emergency Response and Planning
                7.   Audits
                8.   Incident Investigation
                9.   Hot Work
                10.   Prestartup Safety Review
                11.   Mechanical Integrity
                12.   Management of Change
                13.   Contractors
                14.   Trade Secrets
    B.   Models Used
                1.   Vapor Dispersion
                2.   Fires And Explosions
                3.   Quantitative Risk Analysis

VI.   Audit Findings
    A.   PSM History Of The Facility
    B.   General Condition Of The Facility
                1.   Physical Condition
                2.   Maintenance Practices
                3.   Procedures
                4.   Safety Systems
    C.   PSM Elements
                1.   Employee Participation
                2.   Process Safety Information
                3.   Process Hazards Analysis
                4.   Operating Procedures
                5.   Training
                6.   Emergency Response and Planning
                7.   Audits
                8.   Incident Investigation
                9.   Hot Work
                10.   Prestartup Safety Review
                11.   Mechanical Integrity
                12.   Management of Change

It is essential that the auditor thoroughly document all the findings. He or she must take complete notes of:

- Documents examined
- People interviewed — including the scope of the interview
- Physical items examined
- Information retained or copied

LETTER OF CERTIFICATION

The audit should be certified by the auditor. A representative Letter of Certification is provided in Table 15-3.

Table 15-3

Representative Letter Of Certification

The attached audits meet the requirements of the OSHA process safety management standard, 29 CFR 1910.119.

1. The audits were conducted within the three-year program required (paragraph o.1.)

2. One of the auditors (the process safety manager at the facility) was knowledgeable in the process (paragraph o.5.)

3. One of the auditors (from the auditing company) was knowledgeable in audit techniques (paragraph o.5.)

4. The audit included outside field work (paragraph o.7.)

5. Sample sizes were big enough (paragraph o.8.)

LANGUAGE OF THE REPORT

The report should be written in a neutral style. If problems are found, the report should make it clear that these problems are to do with management systems, not specific individuals. (Sometimes the auditor will find a hazardous situation that can be attributed to one individual. In these cases, the auditor should probably communicate that concern verbally.)

The vocabulary used in the report should be careful and objective. Words such as "dangerous" and "willful" are inappropriate. If problems are uncovered, it is better to refer to them as "exceptions", rather than "deficiencies" or "faults."

At all times, the auditor must be careful to distinguish between what was observed and what he deduced. For example, if the auditor noted on two occasions that the facility was carrying out temporary operations without having written procedures for them, the report should state: "Observed that two temporary operations were being carried out without written procedures." The report should not state: "Facility does not have a policy for writing temporary procedures, as evidenced the following two incidents." Even if someone tells the auditor that a policy for writing temporary operating procedures does not exist, the auditor should make it clear that he is reporting that observation. The only conclusion that the auditor can draw from that statement is that that person is not aware of the existence of such a policy, not that such a policy does not exist at all.

The above comments do not mean that there is no place for general conclusions and deductions, particularly in reviews as distinct from audits. However, such conclusions should be identified as such. In the above example, the auditor might write: "There appears to be no policy for writing operating procedures for temporary operations. This conclusion is based on the following facts . . ."

EVALUATE THE FINDINGS/TAKE CORRECTIVE ACTION

Once the report has been issued, the audit is complete. However, the auditor may be asked by the client to use the findings to identify management problems that lie behind the issues that were reported on. To do this, the approach used in Chapter 13 for Incident Investigation can be followed. That is, the elements of process safety are used to structure the analysis of the management systems.

CRITIQUE THE AUDIT

The final step in the audit process is to critique the audit process itself. In particular, the company being audited should tell the auditing company how they perceived the exercise and the value of the findings.

CONCLUSIONS

Even in the best-run facilities, there are always problems. Only by auditing regularly and thoroughly can those problems be identified and corrected. Hence, audits are an essential part of any process safety program.

Chapter 16

QUANTIFICATION OF RISK

INTRODUCTION

∫ Wherever possible, risk should be quantified. Only when this is done can fully rational decisions concerning safety be made and the right priorities be set. Quantification also highlights those safety problems that require immediate attention. As Peter Drucker says, "What gets measured gets done." Quantification is not included in any process safety regulation because doing so would require that the agencies commit themselves to a stated value for "acceptable risk." This would lead to unacceptable political problems.

The topic of risk quantification is an extensive one. What is provided in this chapter is a just a brief overview of the techniques available. Further information is available in the wide variety of books and articles on the topic, some of which are listed in the reference section of this book.

Risk is composed of three elements:

- The hazard
- The consequence of the hazard
- The probability of the hazard happening

A hazard is an intrinsic property of a physical system. It is an event that has the *potential* to cause an accident. (Hence hazards can be neither reduced nor increased; they can only be removed or added.) For example, the potential failure of a pump seal is a hazard, as is the potential for someone falling off a tall structure.

Consequence measures the damage done should the hazard actually occur. If the pump seal does in fact leak, there could be safety consequences (someone is sprayed with chemical), environmental consequences (contamination of the storm water system), and economic consequences (lost production and increased maintenance costs.)

Probability is a measure of how likely it is that the hazard will in fact occur. (The words 'frequency' and 'likelihood' are also used to measure the chance of an accident taking place. As explained later in this chapter, these terms are not completely interchangeable.)

These three elements of risk are usually expressed in a form such as shown in Equation 16-1.

$$\text{Risk}_{\text{Hazard}} = \text{Consequence} * \text{Probability} \dots\dots\dots\dots (16\text{-}1)$$

The use of Equation (16-1) can be illustrated using a simple example.

A tank, T-100, contains a toxic gas. The hazard is defined as a "leak of 20-30 kg of toxic gas from T-100." The consequence of this hazard, should it occur, is that five workers in a nearby control room would be affected with short-term health problems. Based on previous experience with similar processes, it is estimated that this hazard will occur once every ten years. There are three basic options for reducing the risk associated with this hazard.

1. **Remove the hazard.** It may be possible to modify the process so that the toxic gas is not present in T-100, or to use an alternative chemical which is less toxic.

2. **Reduce the consequence.** If a new air conditioning system, which can be isolated from the outside atmosphere, is installed in the control room, this could prevent the toxic gas from affecting the workers quite so severely. Relocating the workers to another control room would be another way of reducing the consequence of the accident.

3. **Reduce the frequency.** A frequency of once every five years is high. If it could be reduced to once every hundred years, say by improving the tank's process controls, then the risk would probably become acceptable.

One term that occurs in the risk management literature is "Probabilistic Risk Assessment", usually abbreviated to PRA. It is a complete, formal, quantified risk assessment. The first PRA, commonly referred to as WASH 1400, was an in-depth analysis of a representative nuclear power station.

A PRA postulates a set of initiating events, such as the loss of electrical power, and then determines the sequence of events and failures that must occur before the major accident in question can take place. A full PRA is a very time-consuming and expensive exercise. However, some of its major tools, particularly fault tree analysis and event tree analysis, can be used in process risk management.

SUBJECTIVE NATURE OF RISK

Although risk is treated in books such as this in a rational and mathematical manner, it must always be recognized that risk is a fundamentally subjective topic; one which cannot be reduced to mathematics only. In other words, the perception of risk can be very different from what the risk actually is. Hence, a mathematical analysis of risk is useful, and it provides a good starting point for any discussion. But it is no more than a starting point; many other factors have to be brought into consideration. (The comments to do with the emotional aspect of risk are concerned mostly with safety and environmental problems. Analyses of the risk associated with economic loss is usually handled more rationally.) Other subjective elements associated with risk include the following:

- *Control.* If someone is in control of a situation, they will usually accept higher risk than someone who is in a dependent position. Hence, a person who drives to an airport may be willing to accept higher risk on the car ride than on the subsequent airplane flight, where he or she is just a passenger. With respect to the process industries, since the community has little or no control of what goes on within the facility, their tolerance for the risk associated with it is usually very low.

- *Familiarity.* Generally, people are more willing to accept risks with which they are familiar than those that are strange or unknown. A person who crosses a busy road every day may be taking a bigger risk (in the mathematical sense) than by living near a chemical plant. But this person understands and accepts the risk of being hit by a car, whereas he or she is apprehensive about the mysterious processes that take place in the chemical plant. Moreover, if someone understands what a chemical is being used for, they are likely to be more tolerant of its use. Hence, the use of chlorine in swimming pools is acceptable, but the use of chlorine in chemical processes may not be.

- *Reputation.* Some chemicals possess a notoriety which may not be fully deserved. For example, hydrogen cyanide — which is a staple of detective stories — is perceived to be extremely dangerous. Yet other substances, such as hydrogen sulfide, which is commonly found in natural gas processing and oil refineries, can be equally toxic.

- *Persons Affected.* It is generally felt that an injury to a member of the public has a bigger impact than the same injury to a worker. This may be unfair and unreasonable, but that it is the way that people think. Furthermore, not all members of the public are perceived to be the same. Injuries to children are usually regarded as being less acceptable than the same injuries to adults.

- *Personal Return.* If a person knows that the risky activity can benefit them, they will be more favorably inclined to accept the risk than if it benefits just strangers. This is at the heart of the NIMBY (Not In My Backyard) problem. People are unwilling to assume the risk associated with a nearby plant when the benefits that it generates are general, but they are the ones shouldering all the risk.

- *Natural Causes.* The public is generally more willing to accept the consequences of natural disasters, such as earthquakes and hurricanes, than man-made disasters, such as chemical spills.

- *Accident History.* People are much less tolerant of facilities which already have a record of safety and environmental problems.

- *Trust.* Some institutions are perceived as being more trustworthy than others. In general, the chemical process industries do not enjoy as high a level of trust when compared with say hospitals and doctors.

- *Number Of People Affected.* Equation (16-1) indicates that there is a direct trade off between consequence and probability. For example, it implies that a hazard that causes one injury per month is equivalent to another hazard that leads to twelve equivalent injuries per year. However, this is not how people view risk. People have more trouble with handling rare, high consequence accidents than they do accepting more frequent, low consequence accidents. For example, in most large American cities, there are about 250 highway fatalities per annum. Although there are many programs in place to reduce this fatality rate, it is basically accepted as a part of life. If, however, there were one airplane crash every year at the any of these city's airports, there would be an outcry. Yet the overall fatality rate is about the same in both cases, i.e. 250 deaths per year. Therefore, Equation (16-1) can be rewritten as follows:

$$\text{Risk}_{\text{Hazard}} = \text{Consequence}^n * \text{Frequency} \dots\dots\dots\dots (16\text{-}2)$$
$$\text{where } n > 1$$

The consequence value has been raised by the power 'n', where n has a value greater than one. Since this value is based on subjective feelings, it is not possible to formally define a value for 'n.' If a value of 1.2 is used, then the adjusted fatality rate for the airplane crash becomes $250^{1.2}$, which is 754. In other words, the *perceived* risk is about three times greater than the objective risk.

It is for this reason that industry, particularly the nuclear power industry, has received so much hostile attention from the public. The ultimate nuclear power plant accident, a core meltdown, has such severe consequences that many members of the public simply will not accept it. The industry has concentrated on reducing the probability of an accident. But this does not eliminate the ultimate worst case, hence many members of the public are not convinced that the industry is safe.

Although the use of quantitative techniques is important because it makes risk analysis more rational, it must be recognized that many elements of risk do not lend themselves to this type of analysis, particularly when it involves understanding the behavior and actions of people. Some differences include:

- *Unpredictability.* Humans are unpredictable, hence they cannot be modeled quantitatively. The consequences of unpredictable behavior may be good or bad, depending on the circumstances, and the actions taken.

- *Improvisation.* The human ability to do something new is unique. It means that humans can identify problems and solutions that had previously not been considered, but it also means that people can do things that are completely unexpected.

- *Emotion.* Human behavior is affected by emotions. Once more, this is impossible to model with mathematical equations.

FREQUENCY-CONSEQUENCE CURVES

Generally, there is an inverse relationship between the frequency with which an accident occurs and the consequences of that accident. This is often expressed as a frequency-consequence curve — an example of which is shown in Figure 16-1. It is sometimes referred to as the Farmer or F-N curve, where F represents the frequency with which a particular accident might be expected to occur, and N represents the consequence in terms of the number of people affected. (Technically, the frequency value, F, actually measures the frequency of exceedance, or the cumulative frequency, i.e. it measures the likelihood with which an accident of consequence N *or greater* will occur.)

Figure 16-1

F-N Curve

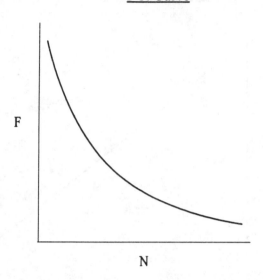

The curve shown in Figure 16-1 is theoretical. In practice, a real F-N "curve" will probably look more like a series of steps, with each step representing an incremental change in frequency and consequence. It is even possible for the curve to have a positive slope over some of its length, i.e. some high consequence scenarios may have a higher likelihood than other, lower consequence scenarios.

Figure 16-1 shows that neither F nor N become zero; both ends of the curve are asymptotic. Consequently, this means that risk can never be zero. Although most analysts have little trouble with this concept, it is important to keep this in mind when talking to those not trained in risk management (for example, members of the public at a hearing to discuss emergency plans.) The correct response to the question "Can [it] happen?" — where [it] refers to some catastrophic accident affecting the public — always has to be "yes."

In practice, F-N curves are usually plotted on log-log paper in order to generate a straight line. Figure 16-2 is the log-log version of Figure 16-1.

Figure 16-2

Log-Log F-N Curve

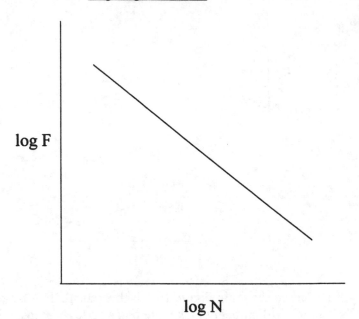

log F

log N

Another factor that enters into risk calculations is cost. Figure 16-3 shows that the reduction in risk achieved by increasing investment falls off asymptotically, i.e. there is less and less return as risk is reduced. Hence, a program that calls for "safety at any price" is disingenuous because costs will eventually rise to the point where the plant will have to be shut down. There are always cost limitations. The critical issue is to make sure that the Cut-Off Point shown in Figure 16-3 lies outside the value of Acceptable Risk.

Figure 16-3

Effect Of Investment On Risk

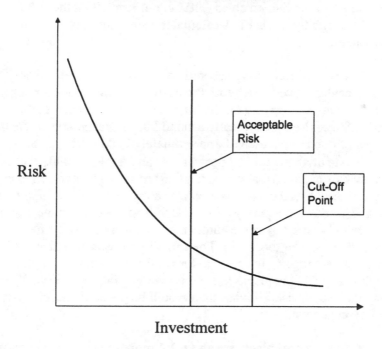

ACCEPTABLE RISK

Because risk can never be zero, it is necessary to determine what level of risk is acceptable. (Some companies set "zero accidents" as a goal. This may provide an excellent target for publicity purposes, but — since risk can never be zero — such a goal is technically unachievable.)

Unfortunately, there is no fundamentally defensible method of determining acceptable risk because there is no objective measure for determining the value of human life, or of the pain and suffering resulting from injury and the loss of loved ones. That is why regulatory agencies will almost never provide a value for acceptable risk; the political implications of doing so will be too controversial. Yet, it is not possible to avoid the question as to what level of risk is acceptable because every decision that is made to go ahead with a project or activity contains within it the implicit acceptance of the risk with which that project is associated. Even everyday decisions, such as crossing a road or taking an airplane ride, are taken with an acceptance of the risk involved (although it is rare that the risk is defined explicitly, or that it is calculated using Equation 16-1 or its equivalent.)

One way of estimating acceptable risk is to compare industry's Fatal Accident Rate (FAR) with the corresponding values for home-based activities or driving on the highways. For example, in the United States there are typically around 50,000 traffic-related deaths per year. For a population of approximately 250 million, this gives a fatality rate of about 2×10^{-4} per year, or about 2 in 10,000. Although society has instituted many programs to try and reduce this rate, there seems to be a broad consensus that it is an acceptable part of normal life and work. A frequency of 1 in 10,000 years for a serious accident has been suggested by some authorities as a working value for acceptable risk (Sutton: 1, 7.) The same figure was also discussed in the Rijnmond report. In practical terms, this means that if a plant employs say 500 people, as has more than one fatality every 20 years, then the level of risk is higher than would be generally accepted by society in general.

In fact, process plants are about 1.4 more times more dangerous than homes. This is very creditable, given that industry routinely handles very hazardous chemicals at extremes of temperature and pressure. Moreover, the process industries are many times safer than most other types of industry.

One way of addressing the question of acceptable risk is not to worry too much about absolute values (which are often very questionable due to inadequacies in the data used.) Instead, it may be more productive to focus on *changes* in risk. Given a choice or risk reduction projects, the one that reduces risk the most is the one to choose, rather than the one which has the highest absolute (calculated) degree of risk.

RISK MATRICES

Once the consequences and frequency of a hazard have been estimated, risk levels can be compared using a risk matrix. This is a two-dimensional matrix, with one side representing consequence and the other side representing frequency or probability. The number of squares on each edge of the matrix typically varies from three to five. Most practitioners find that four or five squares per side give sufficient precision without implying a higher degree of accuracy than can really be justified.

CONSEQUENCE LEVELS

Table 16-1 shows the consequence values that are used to develop a typical risk matrix. The four rows in the matrix represent the seriousness of the consequences associated with a hazard. The columns represent five different types of consequence: worker safety, public safety, environmental, economic and public relations. It provides a rough level of equivalency between the various topics. For example, the persons who developed this matrix decided that a lost-time injury to a worker was equivalent to a minor health effect to a member of the public (both have been assigned a consequence value of "moderate".) This decision is fundamentally arbitrary and subjective. This type of comparison is fundamentally subjective; each facility will have to develop such values reflecting its own circumstances.

Table 16-1

Consequence Categories

Conse-quence Level	Worker Safety	Public Safety	Environ-mental	Economic	Public Relations
1. Low	Minor injury	No injury	Minor impact, remediation limited to $100,000	Less than $10 million	No impact
2. Moderate	Lost-time injury	Minor health effects	Major impact, Remediation in excess of $10 million	Between $10 and $100 million	Slight
3. Severe	Fatality or dis-abling injury	Serious injuries	Remediation in excess of $100 million	Threatens viability of business	Major adverse publicity
4. Very Severe	Multiple fatalities	Fatality	Business threatening impact	Destroys viability of business	National adverse publicity

FREQUENCY LEVELS

Table 16-2 illustrates a choice of frequency values made at this facility. There are five levels, ranging from Extremely Low to Very High. The definition of each of these is shown in the Table.

Table 16-2

Frequency Levels

Frequency Level	Criteria
1. Extremely low	Essentially impossible. (However, very low probability numbers may be misleading due to the presence of unidentified common cause effects.) $<1/1,000,000$ years
2. Low	The event is conceivable. However, there are no records of it having happened, nor have its precursors ever been recorded at this facility. $1/1,000,000$ to $1/10,000$ years.
3. Moderate	At least one of the precursors to the event has occurred at this facility. $1/10,000$ to $1/1,000$ years.
4. High	A review of incident data at similar processes and/or facilities shows that this incident, or one similar to it, has occurred on at least one occasion. $1/1,000$ to $1/100$ years.
5. Very High	The event has already occurred at this facility, and no changes have been made to reduce the probability. $> 1/100$ years.

In Table 16-2, the word "precursor" is used. This refers to the fact that most hazards have protection in the form of safeguards. The precursor is an initiating event that could cause the accident to happen. If this precursor does occur, an accident will not happen because the safeguard will protect the system. Nevertheless, the level of safety has been reduced because the plant is relying on its safeguard systems. Hence risk is increased.

PREPARING THE RISK RANKING MATRIX

Having developed values for consequence and frequency, a Risk Ranking Matrix, such as that shown in Table 16-3 can be developed.

Table 16-3

Risk Ranking Matrix

CONSEQUENCE	FREQUENCY				
	1	2	3	4	5
1	D	D	D	C	C
2	D	C	C	B	B
3	C	C	B	B	B
4	C	B	B	B	A

In this example, the matrix is 5 x 4, corresponding to the number of rows in each of Tables 16-1 and 16-2 respectively. Each square in the matrix is assigned a risk-ranking letter ranging from A to D, where A is the highest. The meaning of each letter is described below.

Level "A"
   Action must be taken, and it must be taken promptly. Money is no object and the option of doing nothing is not acceptable. In practice, it is unusual to come up with Level "A" recommendations on plants that have been operating for some time because it is highly likely that such extreme situations will have already been recognized and addressed.

Level "B"
   The level of risk must be reduced. However, the situation is not as urgent as Level "A". There is time to conduct more detailed analyses and investigations.

Level "C"

The risk should be addressed if the measure is cost-effective. The option of doing nothing is valid.

Level "D"

No action is needed or justified.

The risk associated with a hazard is calculated for each of the consequence types (worker safety, public safety, etc.) The one that delivers the highest overall value of risk ranking is selected. In some cases it may be appropriate to develop a cumulative index. For example, if the risks associated with worker safety and with public safety are each at Level 'B,' it may be decided to give the hazard an overall ranking of 'A.'

RISK RANKING F-N CURVES

F-N curves (described earlier in this chapter) can also be used to define the parameters for acceptable risk. An example is shown in Figure 16-4. Any hazard that falls above the top line is in the area of unacceptable risk. Such hazards must be mitigated in some manner. Hazards that fall below the bottom line are acceptable. Those that lie between the two lines require that some risk reduction work be performed, but it is not critical.

Figure 16-4

Acceptability Of Risk

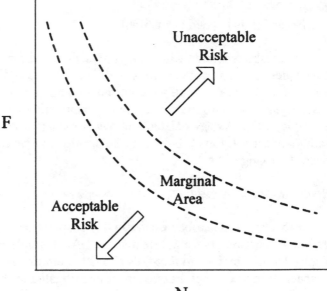

CONSEQUENCE ANALYSIS

Equation 16-1 showed that risk has three components: hazard, consequence and probability. The first of these — hazard identification — was discussed in depth in Chapter 5 (Process Hazards Analysis.) An overview of the next element — consequence analysis — is provided in this section. Generally, consequences fall into one of three broad categories:

- Safety
- Environmental
- Production / Productivity

Most serious accidents to do with process safety are caused either by someone being exposed to a toxic gas or by their being caught in a fire or explosion. Toxic gases are usually of the most concern to the public because it is unusual for a fire or explosion to have effects "over the fence." However, within the facility, fires and explosions represent a serious danger to operators, particularly in those facilities that are handling light hydrocarbons. Regardless of whether a consequence is gas release or fire and explosions, the following factors are important when determining the magnitude of the consequences.

SIZE OF RELEASE

There is almost an infinite range in the sizes of release — ranging from a pinhole leak to sudden and total failure of a major vessel. One plausible scenario is to assume that a gasket fails totally between two adjacent flange bolts. Knowing the gap between the flange faces, the total area of the release can then be determined.

Another approach is to assume that the largest line in the system containing the hazardous chemical fails totally, thus exposing its entire diameter to the atmosphere. Normally, this is not a realistic scenario (although it can happen in cryogenic service if low temperature material enters a carbon steel pipe.)

The problems with estimating release size is one of the reasons the EPA included the worst-case scenario in the RMP regulation; this scenario did at least it put everyone on the same footing, and eliminated debate as to what was "realistic" in a given situation. And the worst case scenario did provide a boundary condition.

The API (American Petroleum Institute) uses a release of 5000 kg (5 metric tons) for the size of the release (Mallett.) The OSHA standard is based on a threshold inventory of 10,000 lb. of flammables.

DURATION OF RELEASE

Neither OSHA nor API discuss the question as to how long a release may go on, i.e. what is its duration. However, EPA does discuss this with regard to its "worst case" scenarios.

A release can vary all the way from a "puff" to a long-duration incident. A puff is what occurs when a small container such as a 55 gallon drum empties itself instantaneously, its contents then vaporize, and then the incident is finished. A long duration release is one that keeps going for many minutes or hours because (a) there is a large inventory of material the can keep the leak supplied, and (b) no action is taken to stop the flow. One of the reasons that the Bhopal accident was so bad was that the dispersion of toxic gas into the community continued for so many hours.

With regard to a flammable gas, there are two release durations to consider. The first is the time that the gas flows before igniting. This period should be as short as possible because, if the cloud becomes large and then detonates, the resulting damage can be devastating, whereas if it ignites straight away the damage is limited to what the fire itself can do. The second release duration is the period of time for which the fire burns before it can be extinguished.

GAS DETECTION SYSTEMS

Gas detectors are used to warn of the release of flammable or toxic gases. They can be used either as area detectors or they can be located to protect specific equipment items. Area detectors are typically spaced 30-40 feet from one another, and are placed around the area that they are protecting. Detectors for equipment items are placed closed to the equipment in the anticipated direction of the release. (If the equipment is protected by water sprays, the detector should be protected from them.) If the gas that is released is heavier than air, the detectors should be placed close to the ground, say at a height of 2 feet. If the material is lighter than air, the detectors are typically placed about 6-8 feet above the anticipated point of release. Often, good advice can be obtained from vendors on issues such as these.

It is particularly important to use gas detectors if the process is contained within an enclosed building because the concentrations can build up rapidly. Similarly, it is a good idea to place oxygen detectors in closed buildings in case there is a release of an inert gas such as nitrogen.

Just as with other operating limits, it is appropriate to have two levels of alarm. The first level will warn that there is a release of the dangerous gas, but — in its present concentrations — it is not hazardous. The second level indicates that the release is large, and emergency action must be taken. For flammable materials, the warning level can be set at 10-30% of the LFL (lower flammable limit.) The emergency level is above 30% of LFL.

The response to these alarms should be worked out in advance. In general, all alarms (even at the lower level) should require the following actions:

- All maintenance work should stop and the maintenance workers should move to a safe location.
- All ignition sources should be extinguished.
- The Emergency Response Team should be notified.

QUALITY OF THE DATA

Poor quality data is a problem that plagues all risk analysis. For example, it is difficult to obtain reliable information as to the effect of toxic chemicals on the human body. It may be possible to conduct tests on laboratory animals. However, most data will be derived by extrapolating health effect information (where the exposure is to low concentrations for long periods of time.) Similarly, reliable data concerning the dispersion characteristics of toxic vapors is rarely available. Occasionally, tests are carried out. For example, release tests were performed for one particularly hazardous gas: hydrogen fluoride. This chemical is widely used in refineries and the electronics industries, and it was important to know how far it would carry when released, and how effective water sprays were at knocking it down. Therefore a series of test releases was carried out in the desert in order to validate the theoretical models. However, in most cases, there is very little raw information, and dispersion analyses are based on theoretical models of the gases.

PERSONNEL PRESENT

The presence of people at a release site is difficult to forecast. In one incident, there was an explosion in an area of the plant which was quite remote from where people normally worked. It would have been be valid to have assumed that the losses associated with the explosion would be just environmental and economic. However, it just so happened that there was a large maintenance crew working in the area on the day of the explosion, and many of them were killed.

By contrast, a refinery had an extremely serious explosion and fire within the heart of its operating units. Although there was a large amount of damage, no one was hurt. Yet a risk analysis of this situation would almost certainly have assumed that someone would be present in the area and that fatalities and/or serious injuries could be assumed, given the magnitude of the accident.

FIRES

There is a good deal of information available on the types of fire that can occur and how to handle them. In particular, both OSHA and the National Fire Protection Association (NFPA) have extensive standards and guidance in this area.

From the point of view of process safety, there are probably two issues to focus on with regard to fires. The first is to do with the presence of sparks or other ignition sources at a facility. If flammable gases are released but do not find an ignition source before they are dispersed, then an accident will have been averted. Since electrical equipment is a likely source of sparks, careful attention is paid to the electrical classification areas. Those areas where a release of gas is possible should be classified such that all electrical equipment will not create an open spark.

The second issue to do with fires and process safety concerns dilution with air. All flammable materials have a concentration range in which they will burn. If the concentration of the material is too high, there is not sufficient oxygen to support a fire if the concentration of the material is too low there is not sufficient fuel to support a fire.

EXPLOSIONS

There are two types of explosion: physical and chemical.

PHYSICAL EXPLOSIONS

A physical explosion is one that takes place without there being any type of chemical reaction, or change in composition of the materials involved. Probably the most common type of physical explosion occurs when an enclosed vessel is heated by an external fire. The gases inside the vessel heat up, the pressure rises, and eventually the vessel explodes (assuming that the relief device is not working.) This scenario is often used as the basis for vessel relief valve sizing. Another type of physical explosion — where water is added to a layer of hot oil in a tank — has already been described (*see* page 158.)

Physical explosions can be just a serious as chemical explosions. For example, in 1997, some high pressure nitrogen vessels in Houston, Texas, exploded, resulting in multiple fatalities. Yet nitrogen is normally seen as being a safe and benign material.

CHEMICAL EXPLOSIONS

A chemical explosion is caused by the energy release from a chemical reaction, usually one that is very rapid. Like fires, chemical explosions are caused by a fuel combining with an oxidant (usually the oxygen in air) to form combustion products with an accompanying release of heat. The basic difference between the two is in the rate at which energy is released. A given amount of material will release the same amount of energy whether it burns or explodes. In an explosion, energy is released so rapidly that a pressure wave (or shock wave) that moves through the air is created. This occurs because the mass of gaseous reaction products cannot expand as rapidly as they are formed.

A chemical explosion can be either a deflagration or a detonation. During a deflagration the pressure increase occurs subsonically, i.e. the pressure front travels more slowly than the speed of sound. This means that there is a chance that the safety systems such as relief valves and interlocks may have enough time to be able to react to mitigate the situation.

A detonation occurs when the shock wave travels faster than the speed of sound. Detonations are generally much more destructive than deflagrations for two reasons. First, because the shock wave is travelling faster than sound the safety systems such as relief valves, rupture disks and interlocks do not have time to respond. The high pressure wave is on them before they know it. Second, the impact of a detonation on the wall of a pressure vessel is such that the metal has no chance to yield; instead, it tends to rupture and to form a large number of small fragments that fly through the air. Deflagrations, on the other hand do give a vessel wall a chance to yield. And, even if the vessel does fail as the result of a deflagration, the pieces created are likely to be much larger than they would be for a detonation. Hence they will not fly through the air and cause as much destruction.

One particularly serious type of accident involving both fire and explosions can occur in large chemical and refining facilities. An explosion of this type is known as a "BLEVE" (Boiling Liquid Expanding Vapor Explosion.) It occurs when there is an external fire around a vessel containing a flammable liquid. The heat of the fire causes the liquid in the vessel to boil. This leads to an increase in internal pressure, which causes the vessel relief valve on the vessel to lift, hence the contents of the vessel to boil away quickly. The flames from the fire then heat up the dry walls of the vessel above the level of the boiling liquid. The resulting high temperature causes the vessel walls to fail. This in turn leads to a massive release of even more flammable materials, thus greatly exacerbating the magnitude of the accident.

BLEVE's are particularly serious when the burning vessel is surrounded by other vessels, as in a tank farm, because there can be a chain reaction of BLEVE-type fires and explosions moving from one vessel to another.

TOXIC GAS RELEASES

Releases of toxic gases are a major safety issue, particularly if the release is sufficiently large to affect members of the public. When analyzing the effects of such releases, there are two parts to the problem. The first is to do with the way in which the gas disperses, and what its concentration is likely to be at various locations. This will depend on factors such as the amount of material released, the size of the release point and the weather conditions at the time. The second important factor concerns the effect of the gas on persons exposed to it. This will depend on the concentration to which they were exposed, the time for which they were exposed, and their general health before the incident.

DISPERSION

The material that is released is usually in one of three forms.

1. Gas or vapor.
2. Liquid below its boiling point.
3. Boiling liquid, resulting in both gas and vapor being dispersed.

When a gas is released, it moves away from the leak source in the direction of the prevailing wind, As it moves, the gas cloud will expand in three dimensions; often developing a cigar-like shape, such as that shown in Figure 16-5.

Figure 16-5

Cloud Isopleths (Plan View)

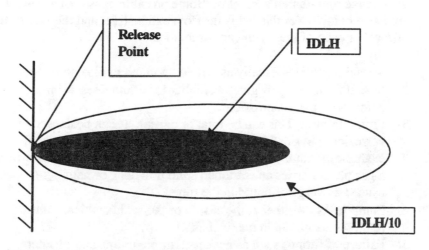

Figure 16-5 shows two isopleths: one for IDLH and the other for IDLH/10 (*see* page 84 for a description of these terms.) Anyone within the boundaries will suffer the health effects associated with that concentration of the vapor. The concentration of gas will be the highest along the center of the cloud. This is why it has its characteristic cigar shape — the center line retains the highest level of concentration. Therefore, those on or near the center line are exposed to a higher concentration than those who are near the edge of the cloud at the isopleth.

Some of the factors that affect the size, shape and dispersion characteristics of the cloud are discussed below.

Vapor Density

Dense gases tend to slump to the ground, light gases rise. For this reason, hydrogen leaks will dissipate very quickly. Even if the hydrogen cloud ignites, it could be well up in the air, thus causing less damage than from an explosion nearer the ground. Dense gases are more dangerous not only because they will be at or near ground level, but because they will be dispersed more slowly.

Size Of The Release

Estimating the size of a release is difficult because there is such a wide range of potential accidents. Some possible causes of a release are listed below. Yet this list is far from exhaustive, and the size of the release for each of these can vary dramatically.

1. Vessel rupture — usually associated with an explosion or fire.
2. Vessel failure — this gives a smaller leak than does a rupture. A common cause is corrosion.
3. Piping failure. This can be total or partial. It can be caused by corrosion, physical damage (such as impact from a vehicle, for example), failure of pipe supports, erosion (this can happen where a gas phase carries an entrained solid powder), or hammer (shock caused by surge of liquid in the pipe.)
4. Pump leaks — either at the casing or the seal (seal leaks are relatively common in many plants.)
5. Failure of fittings such as gaskets, sample points and bleeders.
6. Hose failure, often in association with movement of trucks or barges.
7. Rupture of tanks by a forklift truck.

The problem of selecting an appropriate release size was one of the justifications behind the EPA's decision to require a worst-case scenario for the RMP rule, in which the entire inventory of hazardous chemical is released at one instant. Although such a scenario may not be plausible, it does create a level playing field which can be used for comparing the risks associated with different facilities.

Duration Of The Release

The most important factor in determining release duration is usually the speed with which the operators and/or the emergency response system can react so as to stop the leak. In some cases, it may not be possible to stop it. For example, if a tank nozzle is broken between the tank and the block valve on the line connected to the nozzle, the leak cannot be stopped using normal operational techniques. All that can be done is to remove the contents of the vessel to another safe location as quickly as possible, and/or to quench the exiting gases with a water spray or some other liquid that will safely knock down any escaping vapors.

Atmospheric Conditions

The manner and speed with which the vapor cloud dissipates depends largely on atmospheric conditions. For example, a high wind speed is generally favorable because it will tend to break up the cloud of gas due to its own turbulence. For the same reason, sunny days are better than cool nights because the heat of the sun causes hot (turbulent) air to rise from the ground and dissipate the gas cloud. The direction of the wind is important, particularly if the population around the plant is not evenly distributed.

Sometimes, the release of a toxic gas is accompanied by a fire on the unit. Although the fire creates its own safety and economic consequences, from a dispersion point of view, a fire is "good" because its heat generates turbulence, which in turn breaks up and dilutes the cloud.

Humidity is not generally very important with respect to gas releases. It affects the density of air. Generally, however, this is not an important variable. Although rain can be effective at knocking down water soluble gases, the reality is that — on a statistical basis — it is not usually raining. Therefore, it is not normal to include rain the calculations.

Atmospheric stability is usually divided into six categories from "A" to "F." Class "A" represents the most turbulent conditions (including strong sunlight.) Class "F" represents a still atmosphere at night time.

Ground Roughness

Vapor clouds break up more rapidly when the ground is rough. Roughness can be provided by objects such as trees and buildings. Table 16-4 shows the factors used by the commercial program PHAST for the roughness of the various surfaces.

Table 16-4

Roughness Criteria

SURFACE DESCRIPTION	ROUGHNESS LENGTH
Sea Surface	0.06
Flat Land, Few Trees	0.06
Open Farmland	0.09
Open Countryside	0.11
Woods, Rural Area	0.17
Industrial Site	0.17
Urban Area	0.33

Local Population

Once the model of the plume has been prepared (such as in Figure 16-5), it is overlaid on a map of the community, usually using the most likely wind direction. This map should show areas of human population, including the following:

- Hospitals
- Schools
- Nursing homes
- Shopping centers

- Industrial Facilities
- Major traffic areas and evacuation routes

HEALTH EFFECTS

Health terms such as IDLH (Immediately Damaging To Life and Health) were discussed in Chapter 4 (Process Safety Information.) In practice, the results will be complicated by factors such as the presence of an old persons' home (where the people are more likely to have respiratory problems) or of a prison (where the people inside cannot escape.)

ESTIMATING LIKELIHOOD

DEFINITION OF TERMS

Having identified the hazards and determined their consequences, the final step in a risk analysis is to calculate the probability of the accident occurring. In this context, three words are often used: frequency, probability and likelihood. They have slightly different meanings from one another.

Frequency represents the number of times that an event is likely to occur within a stated period of time. Frequency values have dimensions of inverse time. For example, a process vessel may experience a high pressure situation once every two years. Hence, the frequency of high pressure in this vessel is 0.5 year^{-1}.

Probability is a dimensionless number having a value between 0 and 1. Probability values are generally applied to systems that respond to events, particularly safeguards such as relief valves and interlock systems.

For example, the vessel discussed above may experience a high pressure scenario once very two years. This is a *frequency*. The vessel is protected from explosion by a relief valve which has a *probability* of failure of 1 in 1000, i.e. 0.001. The relief valve does not have a failure "rate." It has a probability of working or not working on demand. The frequency with which the vessel will rupture is determined by multiplying the high pressure rate with the relief valve failure probability. This is 0.5 year^{-1} * 0.001, or 0.0005 year^{-1}, or once in two thousand years.

It is not possible to multiply two frequency numbers because the result would have dimensions of time^{-2}, which is physically meaningless. Physically, this can be interpreted by saying that if two events occur at the same time, then the occurrence of the second event has no meaning because the system has already failed due to the first event. For example, in the case of the pressure vessel discussed in the previous paragraph, there might be two causes of high pressure: a leak through a block valve from another vessel that is at high pressure (with a frequency of once in 10 years or 0.1 yr^{-1}), and overpressure due to pluggage of an outlet vent line during a reaction step (once in 2.5 years, or 0.4 yr^{-1}.)

It is, however, quite in order to add two frequency numbers (the OR Gate in a Fault Tree.) In this example, the sum of the two failure rates provides the overall failure rate of 0.5 yr $^{-1}$.

The third term that is sometimes used is "likelihood." This does not have a precise mathematical definition in the same way that frequency and probability do, instead it is used generically to describe either of the first two terms.

QUANTIFICATION METHODS

The calculation of likelihood can usually be done in one of three ways. The first is using the Pareto Principle, which does not have a theoretical foundation, but which is practical and can be used with minimal effort. The second calculation technique is deterministic analysis. There are many forms of this, including Fault Tree Analysis, Event Tree Analysis and Reliability Block Diagrams. They all share the same basic mathematical approach. The one that is briefly discussed here is Fault Tree Analysis, because this is the one that is most used in risk analysis work. The third way of calculating likelihood is with a stochastic method. Methods that are used include Monte Carlo analysis and Markov Chains. The Monte Carlo method is overviewed here.

PARETO PRINCIPLE

The Pareto principle provides a first estimate for ranking items in terms of risk and reliability. The principle was first developed in the area of economics where it was noted that typically 10-20% of the population own 80-90% of the wealth. For this reason, the principle is sometimes referred to as the 80/20 or the 90/10 rule. The Pareto principle does not have a theoretical basis, but it has a very wide application. For example, in the process industries, it will usually be found that the following apply:

1. 80% of sales come from 20% of the customers.
2. 80% of accidents occur to 20% of the people.
3. 80% of production upsets are caused by 20% of the equipment items.

This principle is sometimes referred to as "Concentrate on the important few — ignore the unimportant many" rule. This means that there is no point trying to improve those items that do not contribute much to the overall problem. If all of the 80% of the "unimportant many" are corrected, the system will still improve by only 20%, whereas if the 20% "important few" are improved by say 50%, the system will improve by 40%.

The Pareto Principle is extremely useful in process risk analysis because it is easy and quick to implement, and because its results are credible. The other techniques are more sophisticated, but they are open to challenge due to low quality of the data and other issues discussed below. By identifying the most important factors that affect safety, the Pareto Principle focuses on the "important few." For example, if most accidents are caused by human error, then the focus of the process safety work should be on administrative issues such as operating procedures and training.

There is no standard way of applying the Pareto Principle. Typically, all the hazards on a facility are listed, and estimates made as to which of them are likely to actually cause accidents. Such estimates will come from hard information, such as accident records, and from analyses, such as those conducted during the PHA's. Then the hazards are listed in order, and action is taken to solve those at the top of the list. An example is provided by Sutton (1: 70.)

DETERMINISTIC METHODS

Deterministic methods for quantifying risk all use the techniques of Boolean algebra. Methods often used are:

- Fault tree analysis
- Event tree analysis
- Reliability block diagrams

Mathematically, all three methods are equivalent. The differences lie in the way in which the problem is structured. Fault trees, for example, focus on the occurrence of undesirable events, whereas block diagrams consider the overall reliability of a system. The method described here is Fault Tree Analysis.

The principles of Fault Tree construction and analysis were described on page 143. What makes the technique so powerful is the ease with which it is possible to prepare quantitative results once the logic has been prepared. A very brief overview of the quantification of fault trees is provided here. The topic is explained in greater detail in a wide variety of books, including those by Sutton: 1, and by Vesely et al. The example used in Chapter 5 (the overflowing tank) will be used here.

THE OR GATE

A simple OR Gate was shown in Chapter 5, Figure 5-5. It is repeated below in Figure 16-6. It shows that the Top Event — Tank Overflows — can occur if either a level gauge fails or the operator fails to notice high level.

Figure 16-6

First Level OR Gate

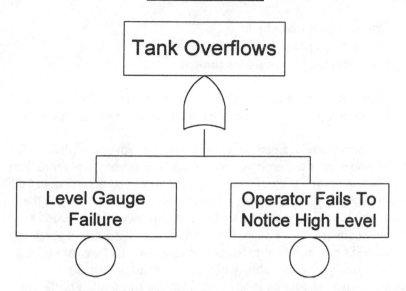

Mathematically, the probabilities of the base events entering the OR Gate are summed. For this example,

$$p \text{ (Tank Overflows)} = p \text{ (Level Gauge Failure)} + p \text{ (Operator Fails To Notice)} \dots\dots\dots (16\text{-}3)$$

where the letter 'p' stands for either probability or frequency. If the frequency of the Level Gauge Failure is say one in 200 days, and if the probability for the operator failing to notice the problem is say one in 20 days of normal operation, then the value for p (Tank Overflows) = 0.005 + 0.05, or 0.055 days $^{-1}$. Hence the tank can be expected to overflow about once in 18 days, and operator error will be responsible for the 91% of the time (illustrating the relevance of the Pareto Principle.)

If it is decided that this frequency is too high, the base event to do with operator error should be developed in more detail. If the Level Gauge Failure problem were to be completely solved, it would still only reduce the overall risk by 9%, therefore there is little incentive to work on it.

Development of the event "Operator Fails To Notice High Level" might show that causes of this error are:

- Operator busy elsewhere
- Operator not properly trained
- Operator read the wrong instrument

The OR Gate is then transformed into an Intermediate Event, with these three new base events entering it. This is shown in Figure 5-6.

Mathematically, Equation (16-3) is slightly over-simplified. It can be seen that if the probabilities of the base events total more than one, then the probability of the Top Event would be greater than 100%, which is physically impossible. The reason for this anomaly is that, if one of the base events occurs, then the occurrence of the second event has no meaning — the system is already in a failed state. Therefore, the second probability value should be ignored. To take care of this mathematically, the probability of the two events occurring simultaneously should be subtracted from the summation of the two probabilities. In the example, therefore, the correct value for p (Tank Overflows) is actually:

$$0.005 + 0.05 - (0.005 * 0.05) = 0.05475.$$

This reduces the top event value by 0.5%, which is not very significant in this case.

THE AND GATE

Mathematically, the AND Gate can be expressed as shown in below, using the example shown in Figure 5-7, which is repeated as Figure 16-7 and is expressed mathematically in Equation 16-4.

Figure 16-7

AND Gate

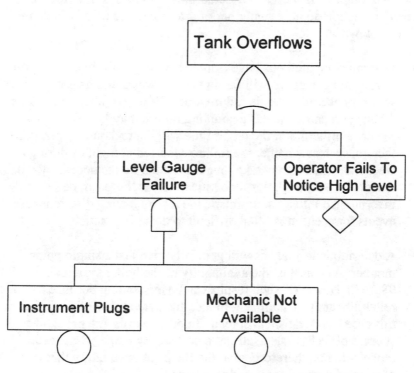

$$p \text{ (Level Gauge Failure)} = p \text{ (Instrument Plugs)} *$$
$$p \text{ (Mechanic Not Available)} \dots\dots\dots\dots\dots\dots\dots (16\text{-}4)$$

Analysis of this situation shows that the initiating event —
Instrument Plugs — has a frequency of once in 20 days, or 0.05 day^{-1}.
The chance of a mechanic not being available is 10% or 0.1, hence the
frequency with which a level gauge failure is expected to occur is
0.005 day^{-1}.

STOCHASTIC METHODS

Although widely used, the deterministic methods described in the previous section do have a number of drawbacks, some of which are listed below.

- Deterministic methods have trouble handling non-linear functions. For example, a pump will fail in various ways, and its downtime will vary according to its failure mode. If a seal fails, it may take 1 hour to repair, whereas a coupling failure may take 6 hours to repair. Furthermore, the repair time will depend on organizational factors, such as whether the failure occurred at night or during a weekend, when it may take longer to round up the necessary repair resources. Although non-linearities such as these can be incorporated into a deterministic model (by having different base events for each), it is often difficult to do so in practice.

- A deterministic analysis will generally give just a single point number, such as the "the availability of the boiler system is 98.3%". However, such results can be misleading because risk and reliability are probabilities. In fact, the availability of the boiler in this case has a statistical spread. In other words, the probability of success of 98.3% has itself got a probability range. Stochastic models do not, therefore, give the flavor of uncertainty that is associated with all real-world systems.

- Communicating the results of a deterministic analysis is difficult, especially to non-specialists. Consequently, the ultimate customer may not be persuaded to take the actions that the analysis suggests.

For these reasons, it is sometimes better to analyze hazardous situations using an entirely different approach known as stochastic simulation. The stochastic method discussed here is Monte Carlo Simulation. (Another well known technique is that of Markov Chains.)

The Monte Carlo method simulates the actual operation of a process on a computer. For example, consider a simple pump set that has two pumps (one on stand-by) that transfers liquid from one tank to another. This is amenable to analysis using a deterministic method such as fault tree analysis. If the failure rates and repair times for the tanks and the pumps are known, an estimate as to the overall system reliability can be made. However, in a real life situation, there are likely to be some non-linear discontinuities in even a simple system such as this. For example:

1. The normal repair time for either pump is 30 minutes. However, the mechanics have to be called out on weekends and holidays. At those times, repair time increases to 2 hours.

2. There is only a limited inventory of spare parts for the pumps. There is normally sufficient material for five repair jobs. Fresh inventory is ordered when there is only enough material for two more jobs. Delivery time for parts is two days.

3. For safety reasons, it is not possible to work on the 'A' pump at the same time as the feed tank. However, it is acceptable work on the 'B' pump and the feed tank at the same time.

    Even small non-linearities such as these are almost impossible to model using a fault tree or equivalent. If this were a real plant situation, with the thousands of complexities that are involved, deterministic models could not be used (or only if some strong assumptions were made.)

    However, this situation presents no problem for the Monte Carlo method. Each event in the system is assigned a probability or failure frequency. This applies to equipment breakdown, repair work, travel time and anything else that represents what actually takes place on the plant. This data is put into the computer model, along with the logic of how the system is put together. Then, a cycle on the computer is set up. The computer simulates plant activity in steps of say 30 minutes. At the end of each computer cycle, a random number generator is applied to each event. Depending on the outcome of this calculation, that event is shown as being in a success or failure mode. This is applied to the model, and the overall system state is evaluated.

    The great advantage of an approach such as this is that it can handle almost any level of complexity and non-linearity because IF statements can be put into each cycle. For example: "IF today is Saturday OR Sunday, maintenance repair times for the pumps are

2 hours." Conditional statements can also be added. For example: "IF there is no inventory of spare parts, do not restore the pump to a working condition."

Another advantage of the stochastic methods is that the results are easy to explain. Non-specialists can grasp what is going on and what they are being told much more easily than they can with say a fault tree because the computer mimics the actual behavior of the plant.

IMPORTANCE RANKING

The chronic weakness of quantitative risk analyses is the quality of the calculated result. There are so many approximations and assumptions in the typical calculation that the quality of the final result is usually questionable. However, this may not be as important as it seems because what is really needed in most cases is not an absolute value for system risk or reliability but a relative ranking of the risks with one another. From this ranking, action items and recommendations can be ordered. Quantitative analysis will eventually result in some type of Pareto ranking, i.e. the "important few" will be separated from the "unimportant many."

What is needed therefore, is a method for systematically ranking base events in a fault tree (or other equivalent method.) Two of the principal means of doing this are the Birnbaum Method and the Fussell-Vesely method.

The idea behind the Birnbaum method is that the reliability or safety of a system is calculated with the base event *in* (i.e. the item is known to be working) and then with the same base event *out* (i.e. the item is defined as not working.) The difference between the two numbers provides a factor for that particular base event. This is the Birnbaum factor, defined as follows:

$$B_{f,i} = p_i (1.0) - p_i (0.0) \dots\dots\dots (16\text{-}5)$$

where p_i represents the system's reliability for the working and non-working cases respectively.

This process of calculating a Birnbaum factor is repeated for each of the other base events. They are then compared with one another. The events with the highest values are the ones that pose the greatest threat to risk and reliability, and should therefore be addressed first.

The Fussell-Vesely method is similar to the Birnbaum method except that the factor for each event is multiplied by the probability of its failure.

The two methods tend to give similar results. The choice between them will depend on the following factors:

- The Birnbaum is more appropriate for those situations where the basic event is known to be fully operational and then fully failed. This represents the situation when allocating maintenance resources when the plant is shut down for a turnaround.

- The Fussell-Vesely approach is more appropriate when analyzing a plant that is already operating. For example, it can be used to prioritize inspection schedules. For this reason, it will be the method of choice for most risk analyses.

FAILURE RATE DATA

All the quantitative methods need basic failure rate data. There are three principal sources of such data:

1. Generic data bases
2. Plant data
3. Expert opinion

GENERIC DATA BASES

There is a variety of reliability data bases that are publicly available. The problem with most of these is that they were developed for non-process industries, particularly the nuclear power business. It then becomes questionable as to how relevant the information from these data bases is in the process industries, where operating conditions and maintenance policies are so different. The book Guidelines for Process Equipment Reliability Data with Data Tables not only provides information for the process industries, but it also identifies many other sources of data.

In spite of the problems associated with generic data, it does provide the analyst with a starting point. As better data become available, the generic data can be replaced. One technique for doing this is through the use of Bayes Theorem.

Sometimes, it will be necessary to estimate the failure rate for an item that never actually has failed. Walker and Lipow provide guidance in this matter. The suggest that the failure rate of $1 / (3*T)$, where T is the life of the item up to that point, is a reasonable value to use for the failure rate.

PLANT DATA

Plant data is usually the best because it is specific to the process being analyzed. However, there can be problems with this type of data.

1. Many plants do not keep a systematic record of equipment failure rates, the failure modes or the repair times. Knowledge of the failure modes is important because most complex items will have various types of failure, each of which is distinct from the others. Averaging them weakens the usefulness of the information. Repair times are also important because, as pointed in the fault tree discussion, overall failure rates are actually more dependent on repair times than on failure intervals.

2. An item's failure rate may no longer be the same as the records would indicate. In particular, if an item has been very unreliable, it may be that it has already been replaced (or the situation causing the problem has been corrected.) Hence the data in the files and computers are misleading.

3. The failure rate for many items of equipment on one site only is simply not high enough to develop a statistically significant data base.

EXPERT OPINION

A valuable source of failure rate data is expert opinion, i.e. the knowledge and experience of the people who work on the plant, particularly the senior operators and maintenance personnel. Naturally, such information is subjective, but it can still be extremely valuable, particularly when combined with hard, objective data. This combination of the two types of data source can be carried using the Bayes method, described in detail by Sutton 1:198.

Some of the cautions about the use of experts that have already been raised need to be carefully considered. In particular, people will tend to place a high probability on an event that has already happened, whereas they will give a low probability to something that they themselves have not witnessed. Obviously, the occurrence or non-occurrence of an event affects the probability of that event occurring again. But, the fact that an event has occurred does not *ipso facto* mean that it has a high probability of occurrence (and *vice versa.*)

COMMON CAUSE EFFECTS

In the Introduction to this chapter, some of the advantages of using quantitative methods to determine risk were listed. However, there are also drawbacks. Possibly the biggest weakness of quantitative methods, particularly in the hands of non-specialists, are overlooked common cause effects.

Consider the AND Gate already discussed in Figure 16-7. It is observed that half of the instrument pluggage problems are caused by solids in the system. However, when solids are present, the mechanic is almost always busy elsewhere. Therefore, at these times, he is not available to repair the instrument. The new tree that develops from this situation is shown in Figure 16-8.

Figure 16-8

Effect Of Common Cause Effect

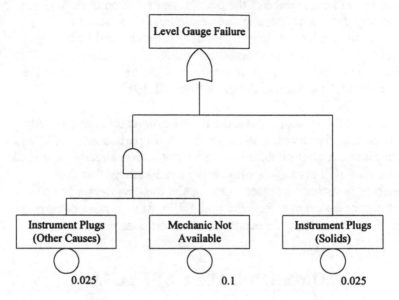

An analysis of this new system shows that the overall failure rate is now 0.0275 day^{-1}. This is more than five times greater than the original value of 0.005 day^{-1}. In other words, the common cause effect has increased the risk by 500%.

It is hard to over-state the importance of common cause effects, particularly when they have not been identified, and so the system safety is much less than an analysis would indicate. This problem is exacerbated by the fact that common cause problems are often "soft", involving human behavior and management systems. Typical common causes are:

- Lack of training for a group of workers
- Insufficient personnel to perform all the work
- Problems with utility systems

CONCLUSIONS

The quantification of risk is a vital part of making process safety management an objective tool that can be used to rationally improve safety and improve profitability. The topic is a very large one, and has only been touched on in this chapter. However, these techniques are thoroughly described in the many books and papers that have been published on the topic.

Chapter 17

IMPLEMENTING PROCESS SAFETY MANAGEMENT

INTRODUCTION

 Up to this point, this book has concentrated on the technical aspects of process safety management. This chapter discusses some of the organizational and managerial issues concerned with the management and implementation of PSM. The central idea presented here is that, although process safety is an on-going, open-ended process, it can nevertheless be divided into a series of finite projects, each of which has a budget, scope of work and schedule. Once one project phase has been completed, the next phase can start.

For example, a company may choose initially to divide its process safety program into three phases or cycles. The focus of the first phase could be regulatory compliance. During this phase it is important to make sure that all the elements have been thoroughly implemented, that any gaps have been fixed, and that the regulatory documentation is complete. The second phase of the program may concentrate on doing what is necessary to achieve a targeted improvement in the safety record. This phase, therefore, may emphasize Incident Investigation and Process Hazards Analysis to find out why accidents are happening and what can be done to reduce their frequency and consequence. The third phase might concern itself with the economic benefits to be obtained from implementing process safety techniques. Therefore, at this stage, the program can incorporate off-site issues that put the plant at economic risk (such as having just one supplier of a critical chemical.)

Regardless of which phase it is in, the organization of process safety can be divided into the following seven steps.

1. Determine Regulatory And Industry Requirements
2. Conduct A Regulatory Audit/Develop A Baseline

3. Determine The Goals
4. Develop A Plan
5. Implement The Plan
6. Audit Progress
7. Start The Next Cycle

Wherever a company may be in this program, it is important always to be ready for an unannouced, hostile audit. There is no telling when an accident might happen, and, if the accident is bad enough, it will be followed by audits from regulators and attorneys representing various plaintiffs. The fact that a facility is in the middle of developing its program is not a valid excuse; an accident can happen at any time, therefore an audit can take place any time.

The practical implications of this understanding are twofold. First, it means that, from the very beginning of the PSM program, it is essential to make sure that all information is properly organized and indexed, so that it can be quickly located and presented to an auditor. This will create a good impression with the auditor, and it will reduce the possibility of not being able to find the information that was asked for.

The concept of organizing around the possibility of a PSM audit is not one that most technical people generally pay much attention to. Unlike other professionals, such as accountants, engineers rarely have their work evaluated by a formal audit. Hence, they do not give the topic much consideration. Furthermore, engineers and other technical people are generally focussed on making their unit is as safe as possible, which means that they are more concerned with doing "the right thing" rather than operating "by the book." Nevertheless, they must plan to meet the letter of the law, as well as its spirit.

A second implication of organizing to meet an audit at all times is that it strongly encourages the top-down approach that was already discussed with regard to operating procedures. Initially, there will be very little detail in the process safety program. But, as it moves through its subsequent cycles, more detail will be added. The key is that, at all times, the program will be complete, and it will not be a collection of miscellaneous documents that may or may not be connected to one another in a coherent manner.

For example, an overview training course for the operators may be put into place. All of the operators will go through this training, and be certified. Once this overview training is complete, management and the operators together can then develop a more detailed course that addresses specific issues in greater depth. This more detailed course can then be given to those operators who are affected. The point is that a complete program is always in place, but there is also always an on-going activity; results are being achieved, but there is always more work to be done to raise standards even further.

Another example is to do with Mechanical Integrity. At the first level of implementation, it may be decided to carry out a simple external visual inspection of all the equipment in the unit. Any deficiencies, such as corroded pipe or leaking instrument air lines can be addressed. Once this overview inspection is complete, management can develop an enhanced thorough program for inspecting and testing equipment using more sophisticated techniques.

The essential point of the top-down approach is that there is always a sense of completion. This is important for the following reasons.

1. If the plant is audited, there is always a complete, coherent, integrated and holistic program to show to the auditor. He or she will probably be impressed with the organization and control of all the PSM activities that this exemplifies.

2. A top-down approach helps to sustain employee morale and enthusiasm. PSM is hard work and can become debilitating for those involved in it; the completion of stages on a regular basis will help create a sense of progress.

3. The initial phases of the program will help identify where additional work and/or a redesign are needed. In addition, as new ideas and suggestions come along (in the spirit of Employee Participation) they can be added on to the existing complete structure and implemented in the next phase.

A practical problem with some process safety programs is that a substantial amount of time and money is spent on them, but —

particularly in the initial phases — there is very little to show for all the work that is being done. (This is particularly true if a top-down approach is followed.) Also, it is not always clear if the final product will meet the facility's requirements. For this reason, it often makes sense to develop some pilot projects at an early stage. A few final products for each element are prepared, and then circulated for comment. This gives everyone a chance to make suggestions.

Some elements of the standard, such as operating procedures and mechanical integrity, lend themselves well to pilot projects. Other elements, such as employee participation and prestartup safety review, are less amenable to this approach.

STEP 1 — DETERMINE REGULATORY AND INDUSTRY REQUIREMENTS

The first step in a process safety program is the identification of those regulations that apply to the facility for which the program is being prepared. If it is located within the United States or its territories, it is likely that it will be covered by OSHA's PSM standard and the EPA Risk Management Program (both of which are described in Chapter 2.) It will also be necessary to determine if any state or local standards apply. Generally, facilities are covered if they handle one or more of the hazardous chemicals identified by the regulation(s) at or above the defined threshold quantities. Therefore, it will be necessary to list all the chemicals that are to be found on site, along with their inventories, and to compare this with the regulatory lists. It is also necessary to determine which non-regulatory standards are to be included. Examples include ISO 9000, API RP 750 and The Chemical Manufacturers Association CAER program.

STEP 2 — CONDUCT A REGULATORY AUDIT/DEVELOP A BASELINE

Having determined which regulations and standards apply to the facility, an audit should be carried out to find out how well the facility currently complies. If the facility has not been audited or reviewed at any time, then this first review is really a baseline audit; it provides management with an understanding where the plant stands, of how much work has been done, and what remains to be done. It also provides the basis for the development of process safety goals.

STEP 3 — DEVELOP THE GOALS

The goals for the process safety program should be as concrete and specific as possible and the level of detail should be tightly defined. If this is not done, a problem often referred to as *scope creep* may take place. As any project progresses, it is tempting to modify the goals to reflect new ideas and to address problems that had not been considered. Such ideas almost always result in an increase in the scope of the project. This is a particularly serious problem with regard to process safety because there is always the feeling that any safety problem has to be addressed and incorporated into the project — not to do so would appear to be irresponsible. The problem with this approach is that adding new items to the project will delay those items that were already scheduled; in effect, there will be a trade-off, and expanding the goals with additional safety targets may also be irresponsible.

STEP 4 — DEVELOP A PLAN

Having established the present status of the facility, and the goals that it needs to meet, the next step is to develop a plan. There are three parts to this:

1. Set up the organization .
2. Estimate what resources are needed.
3. Develop a schedule.

SET UP THE ORGANIZATION

Reporting Structure

Once the process safety program has been designed, the next step is to put in place the organization that is responsible for its implementation. It has been stressed throughout this book that process safety is not a stand-alone, staff-type activity that is conducted independently of the main stream of plant work; it is something to be carried out by everyone in the facility at all times. However, there will also be a need for some staff work. Therefore, on a large facility there will probably have to be one person whose job title is that of Process Safety Coordinator.

Some plants create a Steering Committee consisting of senior people from the major departments in the facility (operations, maintenance, process.) This committee, which should be chaired by the Plant Manager, is responsible for the overall implementation of PSM at that site. It will resolve any problems that are potentially controversial, and it will make decisions on the implementation of recommendations, particularly those recommendations that require significant capital investment. Typically, the Steering Committee will have the following people on it.

- Plant Manager
- Operations/Manufacturing Superintendent
- Maintenance Superintendent
- Safety Supervisor
- Technical Superintendent

Reporting to the Steering Committee are sub-committees, one for each of the major elements of the standard. Each of these sub-committees will be responsible for the development of a detailed policy for their particular element of the standard within the overall company guidelines. Each sub-committee will also be responsible for ensuring that its element is properly implemented and maintained. In practice, this will often mean that they provide assistance and technical support to people in the field.

The chair of each these sub-committees should, where feasible, always be the same person in order to provide continuity. However, the committee membership can constantly rotate, thus giving different people a chance to work the PSM program. Each sub-committee should include at least one employee representative. Not only does this meet the spirit of the Employee Participation element of the regulation, it also helps satisfy legal requirements in this area.

Some of these sub-committees, for example those to do with PHA's and Mechanical Integrity, will be very active and will need substantial resources. Others, such as the Prestartup Safety Review committee, may not have much to do except at turnarounds. Other topics — Trade Secrets, for example — will not usually justify having a sub-committee at all.

The PSM coordinator or facilitator serves as overall project manager. He or she assists the sub-committees and reports to management on overall progress. This person should report directly to the facility Plant Manager. Their responsibilities include the following:

- Finding resources for the people doing the work. The biggest task here is usually to do with people. Process safety requires input from those people who are very skilled in plant operations and maintenance. These people are typically very busy elsewhere, and it can be difficult for them to find the time for PSM activities.

- Training participants in the appropriate process safety techniques. For example, if the operators are to be rotated through a Process Hazards Analysis, it is necessary to provide them with basic training in the technique so that they are able to participate fully and understand what is going on.

- Tracking overall progress.

Figure 17-1 illustrates the type of organization that has just been described.

Figure 17-1

Organization Chart

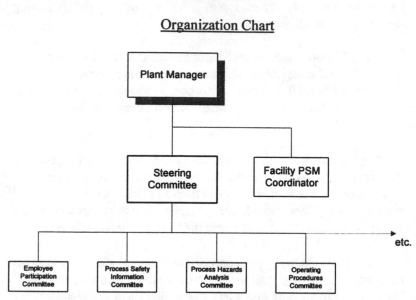

Review Cycle And Signature Authority

Many parts of a process safety program will call for internal and external reviews. Although reviews are necessary, there is a danger that they could become very drawn out, and slow down the whole program. There are two ways of getting around this problem. First, the scope of work should be defined as closely as possible. For example, the level of detail required in the operating procedures should be clarified at the beginning of the project, possibly illustrated using the results of a pilot project. The reviewers should commit to this at the start of the project.

A second way of ensuring that the review process does not bog down the whole process safety project is to make it clear to all reviewers that they are allowed only one review of the relevant document, and that their review should be completed within a certain time frame. If their reply is late then their comments will be ignored. Moreover, once they have submitted their review, further input is not permitted, except to make sure that their first comments were properly understood and incorporated in the final document. (Also, there should always be an opportunity for people to correct factual errors that they find.)

Once the review cycle is complete, it is usually necessary to have the relevant documents signed before they are issued for use in the field. The signing process can also turn out to be a bottleneck, generally for one of two reasons. First, too many people are asked to sign the document. Therefore, it simply takes a long time for it to wend its way through the system. Second, the people who know the unit well, and whose signature is therefore of the highest importance, are also the people who are likely to have the most to do with other projects, or who are busy with day-to-day operations. It can be difficult for them to make sufficient time available to review the procedures and documents with the care and thoroughness that is required.

One way to resolve these problems is to clarify what is meant when a person signs a process safety document. In practice, there are usually just three levels of signature authority that really matter.

1. The first signature is from the person(s) who actually prepared the document that is being reviewed. Their signature states that they have worked to the best of their ability and knowledge, and that the relevant document is, as far as they are able to make it, accurate, complete and useable.

2. The second signature is that of a reviewer who knows the process being described extremely well indeed. Often this will be a supervisor or senior operator. Their signature states that they have reviewed the written procedure or document and that, as far as they can determine, it is comprehensive and accurate.

3. The third signature is that of a manager. It is unlikely that he or she will know the process well enough to comment on the technical content. What their signature states is management systems for writing procedures and other documents are in place, that these management systems address all appropriate regulations and standards, and that these systems were followed during the execution of this assignment.

ESTIMATE RESOURCES NEEDED

The key resources required by a process safety program are money to fund activities such as external audits and equipment inspections, and the time of key personnel to participate in activities such as PHA's and the writing of operating procedures. Although money is likely to be the topic of most concern, the time of key people is often the more important because they are the ones who are also important to all other aspects of plant activity, including operations, maintenance and other types of project work. Hence the process safety schedule will often be developed around the availability of these people.

DEVELOP A SCHEDULE

The non-prescriptive nature of process safety management makes is hard to schedule. This is one reason why it is particularly important to dividing the work into manageable sub-projects, otherwise the task will seem overwhelming. Each phase of the project (and each element within each phase) should be scheduled using normal project management techniques.

STEP 5 — IMPLEMENT THE PLAN

Having developed an organization and having prepared a plan, the first step in implementing a process safety plan — at least on a reasonably large facility — is to develop a Guide which will provide assistance as to the practical implementation of PSM at that site. Issues that could be included in such a Guide are:

- Process safety goals
- Phases of the program
- Means of measuring progress
- Schedules
- Budgets
- Choice of techniques for elements such as PHA's, operating procedures and mechanical integrity
- Management reporting systems

- Review and signature authority
- Organization

If a company has more than one plant, it can develop a corporate level set of PSM procedures. These will then be taken by the plants and modified so as to reflect their particular needs and circumstances.

OPERATING BINDERS

One way of physically organizing the PSM program is to set up a library of 14 binders, one for each element. These binders will contain information associated with that element. They will also contain indexing information that tells the user where other information can be found. As far as possible, the binders and their contents should look the same as one another, and should have similar Tables Of Contents, following a layout such as shown in Table 17-1.

Table 17-1

Representative Table Of Contents For A PSM Binder

1. Introduction
2. Objectives Of The PSM Program
   2.1 Regulations
   2.2 Industry Standards
   2.3 Company Standards
3. Protocols
   3.1 Employee Participation
   3.2 Process Safety Information
   3.3 etc.
4. Administration
   4.1 Equipment Items Covered
   4.2 Personnel
   4.3 Use Of Outside Companies
5. Project Management
   5.1 Phases Of The Program
   5.2 Budget
   5.3 Schedule

In addition to these element-specific binders, there would also be a master binder which will contain the information about the overall progress of the program.

In the spirit of top-down management, all the binders should be complete at all times. Initially, there will be very little detail in them, but they should nevertheless have a section for each part of each element of the standard. The binders should also be physically attractive, and neatly organized. That way, they will make a good impression if the program is audited.

TRACKING CHARTS

The progress of the PSM program can be measured using tables such as Tables 17-2 and 17-3. Table 17-2 shows the progress that has been made in the Training Element of the standard. It indicates that it was estimated that it would require 1260 man-days to complete. So far 525 man-days have been used, so the training element of the project is 42% complete.

In this example, Training has been broken down into two major sections: Initial Training and Certification. (Initial Training is divided into Basic and Detailed.) For each of these, the progress of different sections of the facility is shown. As the project progresses, it will be found that the original estimates as to the level of effort required will require modification. Therefore, the estimate of 1260 man-days will be changed as the project moves forward.

Table 17-2

Training Program Progress

TRAINING						
Level of Effort (Man-Days)						
		1260	525	42%		
	Title	*Number Operators*	*Days/ Operator*	*Required*	*Complete*	*Fraction Complete*
1	**Initial**					
	Basic					
	Utilities	20	3	60	20	33%
	Unit 100	20	3	60	25	42%
	Unit 200	20	3	60	25	42%
	Unit 300	20	3	60	25	42%
	Unit 400	20	3	60	25	42%
	Unit 500	20	3	60	25	42%
	Tank Farm	20	3	60	25	42%
	Detailed					
	Unit 100	20	3	60	25	42%
	Unit 200	20	3	60	25	42%
	Unit 300	20	3	60	25	42%
	Unit 400	20	3	60	25	42%
	Unit 500	20	3	60	25	42%
	Tank Farm	20	10	200	100	50%
2	**Certification**					
	Utilities	20	1	20	0	0%
	Unit 100	20	3	60	25	42%
	Unit 200	20	3	60	25	42%
	Unit 300	20	3	60	25	42%
	Unit 400	20	3	60	25	42%
	Unit 500	20	3	60	25	42%
	Tank Farm	20	1	20	5	25%

A Table such as Table 17-2 is prepared for each of the elements using a spreadsheet. The spreadsheets for each element can be linked to form a master table, such as is shown in Table 17-3. In this example, the total level of effort for the whole program is shown to be 3038 man-days, and that it is at a 31% completion level.

Table 17-3 has a weighting column. This is a measure of how much effort a particular element is expected to take as a fraction of the overall work. In this example, Training is shown as being the most resource-consuming element; it requires 41.5% of the total resources. Incident Investigation, on the other hand, requires only 0.2% of the total resources committed to process safety. Hence, the fact that Incident Investigation is 100% complete is not significant in terms of overall project completion. What is of greater concern is the fact that Operating Procedures, which are estimated to be 18.4% of the total effort, are only 20% complete.

Table 17-3

Overall Progress

	Level of Effort (Man-Days) Total	3038	931	31%	
Element Number	*Title*	*Required*	*Used*	*Fraction Complete*	*Weight*
1	**Employee Participation**	23	11	48%	0.7%
2	**Process Safety Information**	460	96	21%	14.6%
3	**Process Hazards Analysis**	358	72	20%	11.4%
4	**Operating Procedures**	576	114	20%	18.4%
5	**Training**	1260	525	42%	41.5%
6	**Contractors**	40	6	15%	1.3%
7	**Pre-Startup Safety Review**	20	5	25%	0.6%
8	**Mechanical Integrity**	80	29	36%	2.5%
9	**Hot Work Permits**	20	5	25%	0.6%
10	**Management of Change**	130	20	15%	4.1%
11	**Emergency Planning**	41	38	93%	1.3%
12	**Incident Investigation**	5	5	100%	0.2%
13	**Audits**	20	0	0%	0.6%
14	**Trade Secrets**	5	5	100%	0.2%

The order in which the elements will be implemented will depend on the circumstances at each facility. However, the two that should be in place at the beginning are P&ID's and Management of Change (MOC.) Accurate and up-to-date P&ID's are important because so many other aspects of PSM are based on them. They are needed for PHA's, Mechanical Integrity, Operating Procedures and Prestartup Safety Reviews. MOC is vitally important because change is constant. Therefore, as soon as one element is complete, something else will change, possibly affecting the item that was just completed. Therefore, the MOC process must be up and running very early on.

STEP 6 — AUDIT PROGRESS

The importance of auditing has been stressed many times in this book. Simply implementing a process safety program is not enough; management must make sure that its intentions are being properly executed. The topic of auditing is discussed in Chapter 15.

STEP 7 — START THE NEXT CYCLE

Once a cycle of the process safety program has been completed, the next cycle should be started, using the work that has already been done as a basis. This provides an opportunity to move from strictly regulatory and safety issues toward broader topics, such as increased plant reliability and improved product quality.

CONSULTANTS

Companies will often hire consultants to help them with their process safety programs for reasons such as the following.

- Some of the elements of PSM may be new to a company — therefore a consultant can help them get started. For example, in the late 1980's and early 1990's, Process Hazard Analyses (PHA's) were a new concept in most facilities. Hence, a small consulting industry developed to conduct PHA's and to train clients in them. Now that PHA's are generally well understood and widely used, the need for this particular consulting service is not so great (although many of the same people continue to assist with the implementation of the PHA's.)

- A company may be struggling to manage its PSM program. Costs may be out of hand and/or the program may be way behind schedule. A consultant can work with the management team to bring the project back on track.

- Consultants often make excellent auditors. Their expert knowledge of the principles of process safety and of the process safety regulations provides a solid foundation for their findings. In addition, their understanding of the principles of process safety will provide the client management team with insights as to how they need to upgrade the management systems.

- A consultant can provide fresh ideas as to how to perform a task. For example, in Chapter 5 (Process Hazards Analysis) it was pointed out there is a wide variety of techniques that can be used. If a company has become stuck with one method, say the HAZOP technique, a consultant can help them evaluate and use other methods, such as What-If or FMEA.

- A company may require detailed help concerning the interpretation of a regulation or ruling. For example, there is often a question as to how much detail is required in the PSM work. A consultant can provide benchmarks from other companies.

Good consultants possesses the following three traits.

1. They must truly be an expert in the topic that they are advising on.
2. They must be outsiders.
3. They must consult, not do.

TRUE EXPERTISE

True consultants must be true experts. There are many people who know "quite a lot" about a topic, but who are not true experts. In the example quoted above concerning PHA's, by the early 1990's many engineers had become very familiar with the process of leading hazard analyses. This did not, however, qualify them to become true consultants. It simply qualified them to lead hazards analyses. There is analogy here between education and training, as discussed in Chapter 7. Someone who is educated in a topic knows "about principles," whereas someone who is trained knows "how to do it." So it is with consultants and practitioners. Therefore, if a company is looking for a true consultant — as opposed to someone who will simply carry out a task for them — they need to check his or her credentials very carefully. Qualifications to check for include education (particularly advanced degrees), professional publications, presentations and work experience.

Ironically, one of the problems that consultants can run into is that they themselves can become stuck in a rut; they may have trouble accepting that other people's ideas may be as good or even better than theirs. Therefore, it is important to make sure that the consultant is truly up to date, and that he or she is constantly evaluating and testing their own ideas, and abandoning those that are out of date. This being the case, one question that the client company may want to ask a consultant before hiring him or her is: "Which of your opinions and ideas have you changed during the last few years?"

THE CONSULTANT AS OUTSIDER

The consultant should be an outsider. This is important because a consultant may be called upon to present unpalatable truths to management. In many situations the cause of a problem such as a deteriorating safety record is understood by the people at the working level. However, no one feels that they can present "the truth" to management for fear of retribution. (This is not always a management problem, however. The consultant may find that senior management is quite flexible, and willing to adopt new techniques. The resistance may come from supervisors and working-level people who have become entrenched in the current mode of operating.)

A consultant may be able to successfully present bad news for three reasons. First, the worst that the company can do is to terminate his or her contract. Since the consultant usually has other assignments, this loss of work is not so critical as it would be to full-time employee. Second, outsiders are often perceived as being more credible than insiders, even though they present exactly the same facts. (This is why consulting companies themselves sometimes have to hire consultants to tell *them* "the truth". It is also the rationale behind the quotation: "An expert is someone who is more than fifty miles away.") The third advantage of having an outsider present the bad news is that management is not quite sure where to "place" the consultant. Consultants are often perceived as being "above" line employees, particularly if it is suspected that they have the ear of senior management. Therefore, comments from consultants are often treated with a good deal of respect and consideration.

The importance of being an outsider raises a concern about "internal consultants" — a phrase which some might regard as being an oxymoron. If the consultant and the client work for the same organization, sooner or later their chains of command will meet. Hence, neither is truly independent from the other. Furthermore, as their respective careers progress, it is possible that they will find themselves working for or with one another. This knowledge is likely to cloud the objectivity of the client-consultant relationship.

CONSULTANTS — NOT DOERS

The third aspect of consulting is that it is concerned with advising, not doing. A consultant looks at organizational issues, and advises management on how to address them. This is why the end product of most consulting contracts is a report and a presentation to management. In the event that the consultant is asked to implement some of the recommendations contained in the report, he or she has switched roles from being an adviser to a doer.

Good consultants work by generalizing from the specific and then drawing specific conclusions from their generalizations. They go into a situation and investigate the facts of the current situation. From these facts they come up with a general analysis. From that they develop specific recommendations. For example, a consultant may observe that a plant is having a lot near misses with respect to safety. He or she investigates these and finds that many of them can be attributed to the general problem of inadequate training. From this point, the consultant works with the client on developing a training program that is appropriate for this particular situation.

Consultants must possess good client-relations skills. They have to be aware not only of technical issues, but also of the internal company dynamics and politics. Process safety consultants often come out of a technical background — many of them are chemical engineers — and tend to perceive the world as being rational and objective. They may fail to grasp that their clients, like all customers, base many of their decisions on a combination of emotion and fact.

The distinction between "doing" and "consulting" can be frustrating for many consultants. Many of them have had a career in industry, often at quite senior levels. They are used to taking charge and having their ideas put into practice. Hence, the need to persuade rather than command can be very hard for many consultants, particularly when the client chooses to ignore the consultant's recommendations.

Finally, consultants need to grow thick skins. It is almost certain that their ideas and recommendations will be critiqued and criticized. Oftentimes, the people doing the criticizing will be considerably less qualified than the consultant. Also they will have spent less time studying the problem being analyzed, and will probably have motives and agendas of their own. In these situations, the consultant must work as hard as possible to communicate the findings of the analysis to all concerned, but also recognize that the client is paying the bills, and ultimately makes the final decisions. A consultant is an advisor, not a decision-maker.

THE ROLE OF THE CLIENT

The client must realize that the success of the consultant's work will depend largely on the attitude and degree of cooperation provided by the facility employees. In particular, client personnel must try to be open-minded and objective. The consultant has been hired because he or she represents and outside point of view. This means that the findings are likely to upset some people on the client side because old and comfortable ways of doing business will be challenged. The client should try to understand that there may be new and better ways of operating; in particular, everyone should try to avoid using the phrase, "we've always done it that way and it's never been a problem" (with the implication that it never *will* be a problem.)

CONTRACT HELP

In addition to using consultants, a facility may choose to use contract help with many of its process safety activities, particularly those that are labor-intensive, such as writing Operating Procedures or executing the Mechanical Integrity program. The potential drawback to this is that it tends to move the facility away from the philosophy of treating PSM as being a way of life. Contract workers must be considered as part of the team if the process safety program is to be truly successful.

THE FUTURE OF PSM

This book was written as the five-year implementation period allowed by OSHA was coming to an end. From May 1997 onwards, all facilities covered by this standard (i.e. those in the United States) must be in complete compliance with it. During the five-year time period that started in May 1992, many companies have worked very hard on their process safety programs, and are now in a position where they can reasonably claim that they are in regulatory compliance, i.e. were they to be audited by OSHA, not many major deficiencies would be uncovered.

This being the case, these facilities are now examining their programs, and determining how to direct their program from this point.

No one can predict the future, but the following are possible roads forward to be considered.

CHALLENGE THE DESIGN

PSM activities, particularly those for plants that are already in operation, are normally carried out on the assumption that the original design was correct, and that the plant was built and tested according to the appropriate standards. However, such assumptions may not always be true. As companies complete their first-generation of analyses, increasingly they are giving consideration to examining and checking the design bases and the construction standards used when the plant was built.

QUANTIFICATION

Most PSM programs are run by people who have an engineering or technical background. One of the frustrations that such people often experience is that process safety is usually handled largely in a qualitative manner (this was discussed with regard to PHA's, where there is rarely a numerical definition for unacceptable deviations from design or operating intent.)

It is likely that this will change, and that there will be increasing emphasis on the quantitative aspects of process safety. Only when this is done can decisions be made objectively. Moreover, quantification is a prerequisite of the move toward broadening the scope of process safety into the topic of risk management, as discussed below.

TRANSITION TO RISK MANAGEMENT

As facilities move away from regulatory compliance toward achieving objectives in the areas of production, productivity, reliability, environmental performance and safety, they will tend to develop a more holistic risk management approach, using many of the techniques discussed in Chapter 16. Increasingly, those responsible for process safety will have to justify their activities in terms of actual, measurable improvements. Such improvements will be compared to the investment made in PSM in order to determine if the program is indeed making objective sense in terms of the parameters just listed.

No longer will a recommendation be carried out merely because "OSHA says so."

If the transition from process safety to risk management does take place, it is possible that the term Process Safety Management will gradually be phased out, to be replaced with a broader term, such as Risk Management or Loss Prevention.

CONCLUSIONS

Although process safety management is an on-going process that never finishes, it is nevertheless important from a practical point of view, to divide it into a sequence of projects, each of which has clearly defined goals, and which have budgets and schedules. As each phase of the program is finished, the facility can develop new goals for improving safety, improving environmental performance and increasing profitability by ensuring that they are — at all times — in full control of the facility for which they are responsible.

REFERENCES

 There are many books available on the topic of process safety management and on risk management. A few of the key organizations and references are listed below. These are followed by the specific citations.

GENERAL REFERENCES

The *Center for Chemical Process Safety* is sponsored by the American Institute of Chemical Engineers (AIChE). They publish a wide range of books on the topic of PSM. They also provide courses and other information. Information about safety regulations is available from the Occupational Safety & Health Administration (OSHA.) They can be reached at: www.osha.gov.

At various points in this book the Chemical Process Safety Report is quoted. This is a monthly newsletter offering information and analysis on process safety issues. The publisher, Thompson Publishing, can be contacted at www.thompson.com/tpg/enviro/chem/chem.html.

Another excellent source of information is Heinz Bloch's monthly column "HP In Reliability" published in the journal Hydrocarbon Processing. The information in these columns is concerned both with mechanical and systems issues — all to do with keeping a plant running safely and smoothly.

CITATIONS

American Petroleum Institute. Recommended Practice 752. "Management Of Hazards Associated with Location of Process Plant Buildings."

Bloch, Heinz P. "HP In Reliability". Hydrocarbon Processing. April, 1995.

Bradley, Verlon. "Empowering Safety Teams." Texas Chemical Council Safety Seminar: Galveston, TX. (June 13, 1996.)

Brooks, Jr., F.P., "The Mythical Man-Month". New York: Addison-Wesley, (1975)

Carmel, Matthew. Chemical Process Safety Report. (July 1996.)

Center for Chemical Process Safety. Guidelines for Evaluating Process Plant Buildings for External Explosion and Fire

Chemical Manufacturer's Association. Letter to Mr. Roger Clark, OHSA Director. (July 26, 1993)

Chemical Processing "Finding technical help through industry associations." (76) (February 1997.)

Chemical Process Safety Report. (April 1997.)

Chemical Process Safety Reporter. (July 1996.)

Chemical Engineering. "Responsible Care Gains Momentum." (March 17, 1997)

Center for Chemical Process Safety. Guidelines for Evaluating Process Plant Buildings for External Explosion and Fire.

CITGO. A Review Of The Implementation Program For Process Safety Management at CITGO's Corpus Christi Refinery. (1993.)

Elveston, R.. Process Safety Management Conference. Houston, OSHA Region 6. (April 1996.)

Gans, M., D. Kohna and B. Palmer. "Systematize troubleshooting techniques". Chemical Engineering Progress. (April 1991.)

Goodman, Len. "Speed Your Hazard Analysis With The Focused What If?" Chemical Engineering Progress. (July 1996.)

Hoff, R. Document Management. "Software Solutions". (January/February, 1996.)

Hudson, Kevin M. Mechanical Integrity Implementation and Related Process Safety Management. Petro-Safe Conference, Houston, (January 1995.)

Kenney, William F. Process Risk Management Systems. VCH Publishers, Inc.: New York. (1993.)

Kletz, Trevor. Lessons From Disaster: How Organizations Have No Memory And Accidents Recur. Institution of Chemical Engineers, Rugby, England. (1993.)

Knowlton, R.E. A Manual Of Hazard & Operability Studies: The Creative Identification Of Deviations And Disturbances. Chemetics International Company Ltd. Vancouver, British Columbia. (1992.)

Lawley, H.G. Operability studies and hazard analysis. Chemical Engineering Progress. (April 1974.)

Mallett, R. Evaluate the Consequences of Incident. Chemical Engineering Progress. (January 1993.)

OSHA DATA. "It's Confirmed - OSHA Inspectors are Human!". www.oshadata.com/fsoihu.htm. (1997).

Ozog, Henry, P.C. Chatel, C.A. Sabatke. Process Safety Management Auditing. Process Plant Safety Symposium, Houston, Texas. (February, 1992.)

Rekus, John F. OSHA's Lockout-Tagout Standard Mandates Control Of Energy Sources. Occupational Health & Safety. October 1990.

Rijnmond Report No. 87456 Methods For Risk Analysis of the Transportation Of Hazardous Materials In Rijnmond. 1982.

Ross, L.W. "Audit for plant safety." Hydrocarbon Processing. (August 1991.)

Sanders, R.E. Handbook Of Highly Toxic Materials Handling And Management. Chapter 12. ed. S. S. Grossel and D.A. Crowl. New York. Marcel Dekker.

Sanders, R.E. and J.H. Wood. "Don't Leave Plant Safety To Chance. " Chemical Engineering. (February 1991.)

Sutton, Ian S. Process Reliability And Risk Management. New York: Van Nostrand Reinhold, 1992.

---. Writing Operating Procedures For Process Plants. Houston, Texas: Southwestern Books, 1995.

Threet, T.A., Chemical Process Safety Report, August 1992.

Walker, E.I., and M. Lipow. "Estimating The Exponential Failure Rate From Data With No Failure Events." Proceedings 1974 Annual Reliability And Maintainability Symposium.

Weiss, E.H. How to Write a Usable User Manual, ISI Press, 1985

INDEX

D

E

I

L

M